T0372334

Darwinian Heresies

Darwinian Heresies looks at the history of evolutionary thought, breaking through much of the conventional thinking to see whether there are
assumptions or theories that are blinding us to important issues. The collection, which includes some of today's leading historians and philosophers
of science, digs beneath the surface and shows that not all is precisely as it
is too often assumed to be. Covering a wide range of issues starting back in
the eighteenth century, *Darwinian Heresies* brings us up through the time
of Charles Darwin and the *Origin* all the way to the twenty-first century. It is
suggested that Darwin's true roots lie in Germany, not in his native England;
that Russian evolutionism is more significant than many are prepared to
allow; and that the main influence on twentieth-century evolutionary
biology was not Charles Darwin at all but his often-despised contemporary, Herbert Spencer. The collection is guaranteed to interest, to excite, to
infuriate, and to stimulate further work.

Abigail Lustig is a postdoctoral Fellow at the Dibner Institute for the History
of Science and Technology at the Massachusetts Institute of Technology. She
has previously held fellowships at the Max Planck Institute for the History
of Science, Berlin; the Secrétariat National Recherche et Sauvetage, Paris;
and the Universitat Autònoma, Barcelona.

Robert J. Richards is Professor of History and Philosophy and director of
the Fishbein Center for History of Science at the University of Chicago. He
is the author of *Darwin and the Emergence of Evolutionary Theories of Mind
and Behavior* (1987), *The Meaning of Evolution* (1992), and *The Romantic
Conception of Life: Science and Philosophy in the Age of Goethe* (2002).

Michael Ruse is Lucyle T. Werkmeister Professor of Philosophy at Florida
State University. He is the author of many books, including *The Darwinian
Revolution: Science Red in Tooth and Claw* (1979), *Monad to Man: The
Concept of Progress in Evolutionary Biology* (1997), and *Can a Darwinian
Be a Christian? The Relationship between Science and Religion* (Cambridge
University Press, 2000).

Darwinian Heresies

Edited by

ABIGAIL LUSTIG
Massachusetts Institute of Technology

ROBERT J. RICHARDS
University of Chicago

MICHAEL RUSE
Florida State University

CAMBRIDGE
UNIVERSITY PRESS

CAMBRIDGE UNIVERSITY PRESS
Cambridge, New York, Melbourne, Madrid, Cape Town, Singapore,
São Paulo, Delhi, Dubai, Tokyo, Mexico City

Cambridge University Press
The Edinburgh Building, Cambridge CB2 8RU, UK

Published in the United States of America by Cambridge University Press, New York

www.cambridge.org
Information on this title: www.cambridge.org/9780521172684

First published 2004
First paperback edition 2010

A catalogue record for this publication is available from the British Library

Library of Congress Cataloguing in Publication Data

Darwinian heresies / edited by Abigail Lustig, Robert J. Richards, Michael Ruse
p. cm.
Includes bibliographical references (p.).
ISBN 0-521-81516-9
1. Evolution (Biology) – History. I. Lustig, Abigail, 1958– II. Ruse, Michael.
III. Richards, Robert J. (Robert John), 1942–
QH366.2.D342 2004
576.809–dc22 2003069670
ISBN 0 521 81516 9 hardback

ISBN 978-0-521-81516-1 Hardback
ISBN 978-0-521-17268-4 Paperback

Contents

Contributors

Daniel Alexandrov is Professor of Sociology at the European University in St. Petersburg. His area of interest is the sociology and history of science, including the history of sociology and other social sciences. He has worked at the Russian Academy of Sciences and has taught at the University of Chicago and the Georgia Institute of Technology.

Elena Aronova is a researcher at the Institute for the History of Science and Technology, Russian Academy of Sciences, Moscow, where she earned a Ph.D. for her dissertation on twentieth-century immunology. Her research focuses on the history of molecular biology and immunology, Russian experimental biology, and women in Russian science.

Peter J. Bowler is Professor of the History of Science at Queen's University, Belfast. He is a member of the Royal Irish Academy and has taught at universities in Canada, Malaysia, and the United Kingdom. His books include *The Eclipse of Darwinism* (1983), *Theories of Human Evolution* (1986), *The Non-Darwinian Revolution* (1988), *The Mendelian Revolution* (1990), and *Life's Splendid Drama* (1996). He has written several general surveys, including *Evolution: The History of an Idea* (1984) and *The Fontana/Norton History of the Environmental Sciences* (1992). His most recent book is *Reconciling Science and Religion: The Debate in Early Twentieth-Century Britain* (2001).

Abigail Lustig is a postdoctoral Fellow at the Dibner Institute for the History of Science and Technology at the Massachussetts Institute of Technology. She is currently working on a book tentatively titled "Calculated Virtues: Altruism and the Evolution of Society in Modern Biology."

Ronald L. Numbers is Hilldale and William Coleman Professor of the History of Science and Medicine at the University of Wisconsin–Madison. He has written or edited more than two dozen books, including, most recently, *The Creationists* (1992); *Darwinism Comes to America* (1998); *Disseminating Darwinism* (1999), coedited with John Stenhouse; and *When Science and Christianity Meet* (2003), coedited with David Lindberg. He is currently writing a history of science in America and coediting, with David Lindberg, the eight-volume *Cambridge History of Science*. He is a past president of both the History of Science Society and the American Society of Church History.

Robert J. Richards is Professor of History and Philosophy and director of the Fishbein Center for the History of Science at the University of Chicago. He is the author of *Darwin and the Emergence of Evolutionary Theories of Mind and Behavior* (1987), *The Meaning of Evolution* (1992), and *The Romantic Conception of Life: Science and Philosophy in the Age of Goethe* (2002).

Michael Ruse is Professor of Philosophy at Florida State University. He is the author of several books on Darwinism.

Charlotte Sleigh (University of Kent) received her Ph.D. in the history and philosophy of science from the University of Cambridge in 1999 and spent one year as a postdoctoral research Fellow at the University of California at Los Angeles before taking her current post. Her research interests focus on the cultural history of natural history, especially that of insects. Her book *Six Legs Better: A Cultural History of Myrmecology, 1874–1975*, is forthcoming from Johns Hopkins University Press.

Mikael Stenmark is Professor of Philosophy of Religion in the Department of Theology, Uppsala University. He has published papers in the philosophy of religion, the philosophy of science, and environmental ethics and on science-religion issues. He is the author of *Rationality in Science, Religion and Everyday Life: Four Models of Rationality* (1995), *Scientism: Science, Ethics and Religion* (2001), and *Environmental Ethics and Policy-Making* (2002).

CHAPTER ONE

Introduction

Biologists on Crusade

Abigail Lustig

The intellectual landscape of Darwinism for the last 150 years bears a certain resemblance to Germany during the Thirty Years' War. Sects and churches, preachers and dissenters, holy warriors and theocrats vie with each other for the hearts of the faithful and the minds of the unconverted, all too often leaving scorched earth behind.

Such an extravagant metaphor is not much of an overstatement. Accusations of heresy – and equally shameful, imputations of orthodoxy – have been thrown around in the history of evolutionary biology, from within and outside the community of scientists, with reckless abandon. Nor are these terms metaphorical: they are the ones that biologists have used themselves in defense of friends and denigration of foes. Antagonists on all sides of debates about evolutionary biology have wielded the language of holy warriors, declaring crusades to expunge heretics from the domains of biological science. Locutions such as these have become organizing tropes for biologists since the time of Darwin. Yet this aspect of the history of evolutionary theory has – rather surprisingly, in light of the inordinate attention given to evolution's entanglements with religion – usually been ignored.

Why is evolutionary biology so rife with the terms and emotions of organized Western religion? Numerous factors have played a role. Evolutionary biology's emergence from traditions of religious reasoning and writing, into contexts where religious thinking remained prominent; the propensity of evolutionists themselves to paint themselves, ironically or seriously, as dissenters or believers; their tendency to draw, unconsciously or consciously, their scientific frameworks from preexisting religious ones; and their impulse to take

it on themselves to pronounce on issues formerly the domain of religion –
all of these have prompted biologists to armor themselves in the language of
religious combat. We hope that, while this volume may not serve to bring
about the Peace of Westphalia, it may help at least to taxonomize some of the
combatants.

Usage of religiously charged language has a venerable history in evolutionary biology. In the 1910s, the American ant biologist William Morton
Wheeler spoke wryly of his own commission of the eighth and ninth
"deadly sins" in evolutionary theorizing, anathema to the "orthodox behav-
iorists" – anthropomorphism and Lamarckism.[1] Wheeler's German Jesuit
evolutionist entomologist contemporary, Father Erich Wasmann, teasingly
lamented the placement of his own evolutionary works by the great German
Darwinian apostle Ernst Haeckel, "on the index for Monism" for the threat
they posed to "monistic dogmas" asserting the primacy of materialism
and the unity of mind and spirit, which had had, ironically, the opposite effect: "his very denunciation has led no small number of victims into
that snare."[2]

The epithet "apostle" for Haeckel is not misleadingly chosen. Haeckel
played a chief role in the acceptance and substantiation of Darwin's ideas
in Germany, both within scientific discourses – particularly in his work on
marine invertebrates – and in popular culture, which he helped to shape in
best-selling books. Moreover, Haeckel, like E. O. Wilson a century later, explicitly cast science in general, and Darwinism in particular, in the role of
antagonist to and replacement for religion and particularly for Christianity.
In *Monism as Connecting Religion and Science: The Confession of Faith of a
Man of Science* (1895), Haeckel professed a "candid confession of monistic
faith" that he anticipated could replace Christianity.[3] In the mystical and
romantic *Riddle of the Universe at the Close of the Nineteenth Century* (1900),
Haeckel asserted that "what we call the soul is, in my opinion, a natural
phenomenon" and claimed that a monistic view of the universe was tanta-
mount to pantheism, or the idea that divinity inhered in all matter, and was

[1] William Morton Wheeler, "A Study of Some Ant Larvae, with a Consideration of the Origin
and Meaning of Social Habits among Insects," *Proceedings of the American Philosophical Society*
57 (1918): 293–343, at p. 294; William Morton Wheeler, "On Instincts," *Journal of Abnormal
Psychology* 15 (1921): 295–318, at p. 303.

[2] Erich Wasmann, *Modern Biology and the Theory of Evolution*, 3rd ed., trans. A. M. Buchanan
(London: Kegan Paul, 1910), p. xvi.

[3] Ernst Haeckel, *Monism as Connecting Religion and Science: The Confession of Faith of a Man of
Science*, trans. J. Gilchrist (London: A. and C. Black, 1895), p. vii.

"*the world-system of the modern scientist.*"[4] Haeckel's mystic monism was explicitly opposed, rhetorically and substantively, to Christian theology, which
he found scientifically outdated and politically dangerous (in the context of
the German church–state struggles of the late nineteenth century). He hoped
to replace the mealy-mouthed "useless" and "unnatural" love-your-enemies
ethics of Christianity with a monistic morality learned from the "goddess
of truth . . . in the temple of nature," rooted in naturalistic psychology, and
balancing the coequal demands of egoism and altruism.[5]

Herbert Spencer, whose evolutionary philosophy was at least as influential during the late nineteenth century as Darwin's, if not more so, and whom
Haeckel credited with "founding this monistic ethics on a basis of evolution,"[6]
likewise conceived of an ethics that could be detached from transcendental
religious, and particularly Christian, underpinnings. Spencer, however – like
John Stuart Mill with regard to utilitarianism,[7] – prided himself on the asso-
nance between the most highly evolved moral state, to which modern civilized
human beings were tending, and the ethics of pragmatic Anglican Christianity. For Spencer, in fact, the appearance of religious and political authorities
in ages past was a first step on the path that led to the evolution of an absolute
altruism that would require no impetus from outside the individual, being
entirely internalized. The task of the moral scientist was, according to him,
to hasten the "disentanglement" of the latter from the former, as the butterfly
from the chrysalis.[8]

British scientists and theologians of the 1920s and 1930s – combating what
they saw in retrospect as the monolithic materialism of the late Victorian
period, embodied in Spencer and Haeckel – appropriated religious language
to discuss the content and context of science as well. They felt that scientific advances, during the period just before the Modern Synthesis began to
achieve its hegemony, pointed the way to a reconciliation of evolution and
natural theology – usually liberal Anglican but sometimes Catholic – by way

---

[4] Ernst Haeckel, *The Riddle of the Universe at the Close of the Nineteenth Century*, trans. Joseph
McCabe (London: Watts and Co., 1900), pp. 91, 296. See also Chapter 6, this volume.
[5] Haeckel, *Riddle of the Universe*, pp. 362, 345.
[6] Ibid., p. 358.
[7] John Stuart Mill claimed that "[i]n the golden rule of Jesus of Nazareth, we read the complete
spirit of the ethics of utility. To do as you would be done by, and to love your neighbour as
yourself, constitute the ideal perfection of utilitarian morality." See John Stuart Mill, *Utilitarianism*, in *The Philosophy of John Stuart Mill*, ed. Marshall Cohen (New York: Modern Library,
1961), p. 342.
[8] Herbert Spencer, *The Data of Ethics*, 1879. Reprinted together with *Justice* (London: Routledge,
1996), p. 121.

of progressionist evolutionary theories and concepts of "emergence," often
linked to nonmaterialistic physiology, psychology, and comparative sociology. These authors' construction of the Victorians as universally dogmatic
materialists was, of course, factitious, as an equally great diversity of views
on religious and evolutionary issues had been canvassed at all periods since
Darwin.[9] Their sense of being engaged in a great crusade to promulgate a
true view of evolution, however, was evidenced in the titles of their books:
The Basis of Evolutionary Faith (1931); *Landmarks in the Struggle between
Science and Religion* (1925 – taking, despite its title, the opposite side to
Andrew Dickson Carr's famous *History of the Warfare of Science with Theology* of 1896); *The Flight from Reason: A Criticism of the Dogmas of Popular
Science* (1932); *The Gospel of Evolution* (n.d., 1920–1930s).[10] In many of these
works, the metaphorical tables were turned, as Wasmann had done on Haeckel,
to cast mechanistic evolutionists in the role of unthinking "dogmatists"
preaching an unsustainable "gospel."

   The last thirty years have seen an unabashed resurrection of the use of
religiously charged language by participants in evolutionary debates. E. O.
Wilson's announcement of the promulgation of a "New Synthesis," the subtitle of his *Sociobiology* of 1975, helped to catalyze evolutionary biologists
around the revival of thoroughly mechanistic and reductionistic theories of
evolutionary mechanisms, particularly W. D. Hamilton's inclusive fitness or
kin selection theory. A number of these biologists – among them Wilson and
Hamilton themselves, and including the likes of Richard Dawkins, Richard
Alexander, and Robert Trivers – asserted that their theories of the origins of sociality and social behavior, including human sociality and behavior, had grave
implications both for the origins of human morality and for the historical
appearance and development of religion.[11] Several have further asserted, like
Haeckel, that evolutionary biology, in one form or another, is slated to replace
religion in its social functions as well. A number of these biologists have confessed to "conversion experiences" of one kind or another, in which a youthful

[9] See Chapter 3, this volume.
[10] Floyd E. Hamilton, *The Basis of Evolutionary Faith: A Critique of the Theory of Evolution*
   (London: James Clarke, 1931); James Young Simpson, *Landmarks in the Struggle between
   Science and Religion* (London: Hodder and Stoughton, 1925); Arnold Lunn, *The Flight from
   Reason: A Criticism of the Dogmas of Popular Science* (London: Eyre and Spottiswood, 1932);
   J. A. Thomson, *The Gospel of Evolution* (London: George Newnes, n.d.).
[11] See Richard Dawkins, *The Blind Watchmaker* (New York: Norton, 1986); Richard Alexander,
   Darwinism and Human Affairs (Seattle: University of Washington Press, 1979); Robert L.
   Trivers, "The Evolution of a Sense of Fairness," in *Absolute Values and the Creation of the New
   World: Proceedings of the Eleventh International Conference on the Unity of the Sciences* (New
   York: International Cultural Foundation Press, 1983), pp. 1189–1208.

faith in organized religion came to be replaced by modern neo-Darwinism – this narrative too harks back to Haeckel.[12]

Partly as a result of these perceived challenges to religion, and partly as a consequence of quarrels within evolutionary biology, Darwin's modern apostles have been much given to the invocation of religious language in their writing in order to defend themselves and to anathematize their scientific and cultural opponents. George C. Williams, in *Adaptation and Natural Selection* (1966), a seminal work of the new synthesis, argued that the "ground rule [of Darwinism] – or perhaps *doctrine* would be a better term – is that adaptation is a special and onerous concept that should be used only where it is really necessary" – a teaching seldom heeded. Williams concluded the book by asserting, with deliberate provocation, that although the strict modern theory of natural selection "may not, in any absolute or permanent sense, represent the truth . . . I am convinced that it is the light and the way." L. B. Slobodkin, in reviewing *Adaptation and Natural Selection* for the *Quarterly Review of Biology*, picked up at once on the tenor of Williams's crusade. "Williams has written," he observed, "a polemical tract against what he considers to be heresies and deviation in Neo-Darwinian orthodoxy." He continued wryly: "When heresy is being sought out, I am always slightly nervous until I can analyze precisely who the heretics are. Perhaps I, too, am a heretic."[13] Williams's chief heretic, notoriously, was not Slobodkin but Vero Wynne-Edwards and his notion of group selection. David Sloan Wilson noted tartly that by the 1990s group selection had come to be "treated as such a heresy that the only thing to learn about it is 'Just say no.'"[14]

The advent of sociobiology has provided the most vitriolic accusations of heresy and orthodoxy in modern biology. Mary Jane West Eberhard, in a prominent review of *Sociobiology* for the *Quarterly Review of Biology* in 1976, cast sociobiology's genesis in mythic terms by rewriting it as a parable:

[T]here was one small group [of biologists] *without a name. They went about dressed in the castoff clothing of the titled sciences, and often failed to recognize each other, even when they hurried along the same paths. So they suffered greatly. Sometimes they had to learn to collect birds or identify ants in order to get jobs. Then one day there rose up a man*

[12] E. O. Wilson, *Naturalist* (Washington, DC: Island Press, 1994). See also Chapters 6 and 9, this volume.

[13] L. B. Slobodkin, "The Light and the Way in Evolution [review of G. C. Williams, *Adaptation and Natural Selection*]," *Quarterly Review of Biology* 41 (1966): 191–4, at p. 191.

[14] D. S. Wilson, "Introduction: Multilevel Selection Theory Comes of Age," *American Naturalist* 150, supp. (1997): S1–S4, at p. S2.

from among them. He had been called Entomologist, Ecologist, and even Biochemist. But that was not enough. All grew quiet as he raised his golden pen: "There shall be a new science," he said, "and it shall be called SOCIOBIOLOGY."

And the opening sentence of her review left no doubt of the cultural valences she intended to invoke – sociobiology's founder as benevolent God the Father: "Edward Osborne Wilson, the kindly bespectacled father of sociobiology, has assumed god-like powers with this book."[15] W. D. Hamilton – a darker, more pessimistic person – wrote with a certain self-congratulatory relish of the "heresy" he had unleashed on a complacent world, which "for the re-slanted spiritual descendants of the prim Victorians [remains] quite paralysing": the idea that inclusive fitness implied that members of a group "need and are expected to evolve a degree of xenophobia" and, in general, that the selfishness of genes implied the innate selfishness of people. A scientist had to be "tough" – by implication, tougher than any religious adherent could be – if he were to contemplate such painful truths.[16]

Participants in sociobiological controversies have been particularly fond of portraying themselves as martyrs – Galileo or Giordano Bruno by choice – condemned by the Catholic Church. Who plays which role, of course, depends upon the martyr's scientific and political position. Wilson, beset just after the publication of *Sociobiology* by controversy sparked by Harvard's Science for the People Sociobiology Study Group, compared himself to the Swiss theologian Hans Küng, "facing the fury of the theologians" for his liberal Vatican II views.[17] Alexander Morin made the category of "heresy" central to a 1979 analysis of the controversies, "Revelation and Heresy in Sociobiology," in *Science, Technology and Human Values*: "The attempt to 'biologicize' the social sciences is resisted with the same ferocity that the Roman Church brought to bear on the Albigensians."[18] On the opposite side, the sometime Science for the People member Stephen Jay Gould deplored in 1979 the "expanding orthodoxy" of the modern synthetic theory, contrasting it with a "Darwinism . . . sufficiently broad and variously defined to include a multitude of truths

[15] Mary Jane West Eberhard, "Born: Sociobiology [review of E. O. Wilson, *Sociobiology*]," *Quarterly Review of Biology* 51 (1976): 89–92, at p. 89.
[16] W. D. Hamilton, *Narrow Roads of Gene Land: Volume 1. Evolution and Social Behaviour* (Oxford: W. H. Freeman, 1996), pp. 188–9.
[17] In W. R. Albury, "Politics and Rhetoric in the Sociobiology Debate," *Social Studies of Science* 10 (1980): 519–36, at p. 524. Wilson, "What Is Sociobiology?" in Gregory et al., eds., *Sociobiology and Human Nature*, pp. 1–12.
[18] Alexander J. Morin, "Revelation and Heresy in Sociobiology: A Review Essay," *Science, Technology & Human Values* 4, no. 27 (Spring 1979): 24–35, at p. 32.

and sins."[19] Later in life, Gould devoted numerous essays and a book, *Rocks of Ages*, to both celebrating the connections between and policing the boundaries of science and religion; he also never tired of contrasting his own evolutionary views, which were outside the central stream of the new synthesis, with the latter's suffocating "orthodoxy."[20]

Why is a modern science so riven with accusations reminiscent of the Spanish Inquisition? In great part, it is a product of the fact that evolutionary biology emerged within Western, largely Christian societies, tied at its birth to traditions of natural theology. The late nineteenth century was, moreover, a period of struggle over the implications of secularizing worldviews, driven not only by biology but also by anthropology, sociology, and biblical criticism. Evolution offered origin narratives that both echoed and threatened Christianity's, as in Darwin's evocation of the great Tree of Life of phylogenetic descent, given indelible visual form by Haeckel. The *Kulturkampf* of the 1870s–1890s between the newly unified German state and the Catholic Church helped to inflect German evolutionary biology with the crusading tone so characteristic of Haeckel and Wasmann.

The American context, which has been the scene of so much of the most vituperative counteraccusations of orthodoxy and heresy during the late twentieth century, has been particularly prone to this evangelizing mixture of the languages of evolution and religion, for two reasons. The first is the characteristically American history, continuing through the twentieth century, of religious fervor and revivalism, leading to anti-evolution outbreaks such as the carefully staged Scopes "monkey trial" of the 1920s. The rise of Protestant fundamentalist denominations that insisted on biblical literalism – in contrast to the long accomodationist intellectual traditions of Catholicism and Anglicanism – led believers of these sects – as, for example, the Seventh-Day Adventists – even to challenge evolutionists on their home ground.[21] The second factor virtually guaranteeing conflict between fundamentalist Christianity and evolutionists, which had the effect of causing evolutionists to solidify their own ranks and to feel themselves besieged by hostile Christianity – not the metaphorical Inquisitions of sociobiology's critics but a literal war for souls – is the curious fact that, unlike the situation in most modern Western democracies, control over American school curricula is exercised exclusively

---

[19] Stephen Jay Gould, "Is a New and General Theory of Evolution Emerging?" *Paleobiology* 6 (1980): 119–30, at p. 119.
[20] Stephen Jay Gould, *Rocks of Ages: Science and Religion in the Fullness of Life* (New York: Ballantine, 1999). See also Chapter 9, this volume.
[21] See Chapter 5, this volume.

at town, county, and state levels rather than through centralized national oversight. In practice, this has meant that, while the question of teaching evolution in European schools was settled decades ago, American biologists are still called out time and time again to defend themselves and their science in local and state school disputes. This has had the effect of encouraging them to defend their own orthodoxies – in this case, the fact of evolution and the theory of natural selection as its explanation – and to regard with suspicion any member of their own ranks who appears to present a weak flank to the enemy.

In other situations less charged with general cultural religious fervor, the language of heresy and orthodoxy in evolution has been changed or muted. Soviet biologists, of course, had to be careful in seeming to adhere to a different, aggressively secular, set of orthodoxies – Marxism-Leninism. Caught between intellectual orthodoxies, Soviet biologists, particularly during the period of Lysenko's hegemony, risked being placed in an awkward position in which "to be an orthodox geneticist was equal to being a political heretic."[22] Here it was questions of the political rather than the spiritual authority of knowledge that dictated evolutionists' work and rhetoric.

Likewise, in the first enthusiasm for evolutionary ideas in non-Christian Japan, imported through the works of Darwin and Herbert Spencer, it was political rather than religious valences that were invoked by evolutionists, as Meiji reformers used evolutionary ideas to subvert the conceptions of nature that the Tokugawa shogunate had used to bolster its own claims to political legitimacy. The resonances and conflicts present in the West between evolutionary origin narratives and those provided by Christianity were quite absent in Japan, allowing thinkers there to turn evolutionary narratives to distinctly national ends, in a society historically lacking strong or unified religious authorities.[23]

Finally, another reason for the strong resonances between religious and scientific disputation in the case of evolutionary biology is often overlooked: both the Western monotheistic religions and evolutionary biology are to a strong, distinctive, and somewhat anomalous degree text-based. Evolutionary biologists use texts, particularly in the form of books, to a far greater degree than other modern sciences. Of the texts of lasting importance, Darwin's *Origin of Species* holds the preeminent place. On a rough count of the *Science Citation Index*, it has been cited a couple of thousand times in the period

---

[22] See Chapter 3, this volume.
[23] See Julia Thomas, *Reconfiguring Modernity: Concepts of Nature in Japanese Political Ideology* (Berkeley and Los Angeles: University of California Press, 2001), especially Chapters 4 and 5.

and publications covered by the index, mostly by biologists – approximately 500 times since 1996 alone. By comparison, Newton's *Principia Mathematica*, a work of immense importance in the history of mathematics and physics, is cited an order of magnitude less often – a few hundred times – and perhaps a third or more of those citations come from articles on the history of science, philosophy and the history of philosophy, or other disciplines outside physics and mathematics; nor are student physicists expected to read the *Principia*. But particularly since the architects of the Modern Synthesis made reference to Darwin part of their project of returning natural selection to centrality as an evolutionary mechanism,[24] evolutionary biologists have tended to use Darwin in two ways: either to prove that he agreed with their argument, by pointing to passages in the *Origin* or elsewhere in Darwin's works that in their reading foreshadow their own conclusions; or to argue that he *would* have agreed with them, had he had the benefit of information that he lacked but that is available to the modern scientist. To this end, Michael Ghiselin, Stephen Jay Gould, Richard Lewontin, Ernst Mayr, Gareth Nelson, E. O. Wilson, and many many others pore over the text of the *Origin*, the *Descent of Man*, and other works with the assiduity of Talmudic scholars.

A representative example may be found in a quarrel in 1974 between the practicing systematists Ernst Mayr and Gareth Nelson, in taxonomic journals, over whether Darwin's philosophy of classification in the *Origin* agreed with Mayr's own or with Willi Hennig's cladistics. While Nelson acknowledged that "[i]t may be that Darwin's remarks will ever remain ambiguous by modern standards; that may be their virtue (their capacity to be reinterpreted anew)," he nevertheless decided, in the next sentence, "Perhaps so, but perhaps not." Mayr's and Nelson's disputation over texts took on, yet again, the trappings of religious conflict. "I consider historically inaccurate," Nelson groused, "Mayr's repeated assertions that 'evolutionary taxonomy' is orthodox Darwinism and that, consequently, 'cladism' – as Mayr uses the word, is some recent, more or less rootless, heresy."[25] To be able to evoke Darwin's textual authority for one's own position is here a way not only of seizing the scientific high ground, but also of recapturing the idea of "orthodoxy" as desirable: the community of true believers descended from the patriarch – a contrast to the usage of

---

[24] See Vassiliki Betty Smocovitis, "The 1959 Darwin Centennial in America," *Osiris* 14 (1999): 274–323; and Chapter 7, this volume.

[25] Ernst Mayr, "Cladistic Analysis or Cladistic Classification?," *Zeitschrift für Zoologische Systematik und Evolutionsforschung* 12 (1974): 94–128; Gareth Nelson, "Darwin-Hennig Classification: A Reply to Ernst Mayr," *Systematic Zoology* 23 (1974): 452–8, at p. 453.

sociobiologists, who have delighted in portraying themselves as "heretics" persecuted by a world of hostile Marxist and anthropological orthodoxies.

These shifting uses of "Darwinian heresy" and "orthodoxy" reflect also the historical fact that the content of what is called "Darwinism," and therefore the accepted canon of texts and dogmas, has been ceaselessly shifting over the last 150 years. Consider just a few of the issues that have drifted in and out of favor, often without regard to their presence or absence in Darwin's own works. Foremost among these is Lamarckism, which Wheeler called the "ninth deadly sin" of orthodox Darwinism. Notoriously, no other mechanism in evolutionary theory's history has come in for the opprobrium of the in-heritance of acquired characters, even before the fiasco of Lysenkoism. The inheritance of acquired characters was, as is well known, an essential evolu-tionary mechanism in Darwin's own writings, particularly in relation to the evolution of instinct and of social behaviors, but in the Weismannism that wielded great power among theoretical biologists early in the twentieth cen-tury, and which was often called at the time "neo-Darwinism" (no term except "Darwinism" itself has been so often reinvented in the history of evolution as "neo-Darwinism"), natural selection precluded all other mechanisms. A self-described "antiquated" natural historian like Wheeler could thus feel himself on the defensive against it and indeed take pains to denigrate Darwin's charac-ter and achievements, in a manner unthinkable for an orthodox evolutionary biologist, of whatever sect, today.[26]

Lamarckism's tortured history reached an apogee with the Lysenko affair in the Soviet Union during the 1940s and 1950s. While the struggle there was couched in terms not of religious but rather of Marxist-Leninist orthodoxies, the conflicting demands of politics and scientific fidelity could put evolu-tionists in very awkward positions indeed. The evolutionary entomologist Georgii Shaposhnikov, for example – an aphid taxonomist by training who had performed a series of meticulous experiments during the 1950s demon-strating the rapid speciation of asexual lineages of aphids introduced onto new plant hosts (their parthenogenetic reproduction precluding the action of natural selection on new variants) – was forced to walk a careful line in interpreting his results against a series of changing orthodoxies, political and scientific, over five decades. His experiments appeared definitively to demon-strate Lamarckian speciation; yet in order to avoid being pulled into the witch hunts taking place over Lysenkoism, he refrained from publicly interpreting his own results until the 1980s, lest he be excommunicated (or worse, given

[26] William Morton Wheeler, "The Ant-Colony as an Organism," *Journal of Morphology* 22 (1911): 307–25, at p. 307. On Wheeler's downplaying of Darwin, see Chapter 8, this volume.

the penalties applied to Soviet scientists who fell afoul of politics) by one side or the other.

And yet historically, Lamarckism has also been the most useful of heresies; there are times, as with Shaposhnikov, when it seems that no other mechanism can explain the facts, and biologists have resorted over and over again to reinventing the so-called "Baldwin effect" (in which chance genetic variations arise to fit useful variations induced by the environment) in order to invoke Lamarckism's power.[27]

Does "orthodox" Darwinism include a notion of teleology or progress? Biologists and historians of science have quarreled incessantly over Darwin's *ipsissima verba*; however, all this pilpul has reached no set conclusion as to whether the patriarch himself believed in the notion of progress as an integral part of his overall evolutionary views. As he did so often, Darwin seems on this issue to have had his cake and eaten it too.[28] That most other evolutionists of Darwin's time and subsequently were progressionists is beyond doubt, and some observers have sought to place primary blame for this fault – if fault it is – on Herbert Spencer and Ernst Haeckel, reserving purity of ateleological intent for Darwin himself.[29] Whether this faith in overall progress is justified is a surprisingly open question – surprisingly few biologists have sought to test the notion in any rigorous way.[30] Moreover, where biologists fall out on a spectrum of faith in progress versus belief in ateleology seems to have little correlation with their views on other orthodoxies, including the primacy of natural selection or the existence of God: both William Morton Wheeler and W. D. Hamilton, for example, were dysteleologists, believing in the possibility or even probability of the devolution and degeneration of the human species, while their respective contemporaries and allies, Auguste Forel and E. O. Wilson, have been sunny optimists confident in the possibility

---

[27] On the history of the Baldwin effect, see Robert J. Richards, *Darwin and the Emergence of Evolutionary Theories of Mind and Behavior* (Chicago: University of Chicago Press, 1987). For an example of its modern invocation as an ad hoc evolutionary mechanism, see E. O. Wilson, *Sociobiology: The New Synthesis* (Cambridge: Harvard University Press, 1975) – on the origins of religious instincts in human beings, see especially pp. 559–62, discussed by Lustig in Chapter 4, this volume.

[28] See Robert J. Richards, *The Meaning of Evolution: The Morphological Construction and Ideological Reconstruction of Darwin's Theory* (Chicago: University of Chicago Press, 1992), and Chapter 1, this volume.

[29] See Chapter 7, this volume, and Michael Ruse, *Monad to Man: The Conception of Progress in Evolutionary Biology* (Cambridge, MA: Harvard University Press, 1996).

[30] For someone who has attempted to test the idea of evolutionary progress, see Daniel W. McShea, "Metazoan Complexity and Evolution: Is There a Trend?," *Evolution* 50 (1996): 477–92.

of reaching "pinnacles of social evolution" (as Wilson put it);[31] and those incessant antagonists on so many issues, the great atheist Richard Dawkins and the theistic accommodationist Stephen Jay Gould, have both insisted on the meaninglessness of any such criterion as "progress."

Likewise with the tortured question – again so close to questions of orthodoxy, heresy, and evolution's relationships with religion – of whether evolutionary biology provides any explanation of the origins of morality that would warrant using it to found a normative ethics. Attempts to stake out and defend positions on this issue, which have inevitably tended to echo and indeed to bring on further conflicts with the modern monotheistic religions (it should be remembered that morality and normative ethics are not characteristic of all, or perhaps even of most, religious systems), have led to some of the bitterest acrimony among evolutionists and critics. Those who think that evolution teaches us valuable lessons about the origins of human morality nonetheless differ in sectarian ways as to what we should do about it: Is the fact that our morality *is* evolved sufficient to found a normative ethics? Or is that morality so hopelessly limited that our only hope for a genuine ethics is to defy our biology? Can and should human feelings of reverence be detached from religious objects and replaced by the narratives of science and biology? Or are religious sentiments valuable in themselves? Can religious forms of thinking about the meaning of life be reconciled with scientific analyses of it? Evolutionists show no sign of settling these quarrels, which have already occupied them for 150 years.[32]

All of these disputes, impinging as they do on the borders between biology, philosophy, and metaphysics, point to another set of disagreements, often masked. What is evolutionary biology *for?* Is its purpose to explain the shape of nature? Or is it to describe history, whether natural or human or both? Or is its ultimate purpose to explain human nature, and if so, why? – for mere academic interest, or in order to do something about it, whether by changing our biology or by altering our society or culture in the light of evolution's teachings? Does what evolution tells us give a different meaning to our lives, and should it? And in either case, what is its proper relation to other systems of thought, such as religion, that do the same work? With these questions unsettled, it seems unlikely that evolutionary biologists' doctrinal disputations will soon evaporate.

Opinions on all of these questions have shifted with changes in biological information and with fashions in theories, including shifts in the kinds of questions that evolutionary biologists are interested in answering at one period

---

[31] Wilson, *Sociobiology*, Chapter 18.   [32] See Chapters 3, 4, 5, and 9, this volume.

or another: the aspects of the origins of sociality that interested biologists at the fin-de-siècle, for example, were quite different from those that have interested sociobiologists. Concomitant with these historical variations have been variations in what is seen to constitute a Darwinian heresy – the more so since the content of the very word "Darwinism" has never been the subject of unanimous ruling. This has been true for both biologists and those who observe them, and the reader of this volume will find as great a spectrum of opinions about, and epistemological commitments to, the subjects briefly delineated in this introduction in evidence among the authors here as among their subjects. This befits a subject whose implications have been, since the idea that life on Earth had a material origin and history was first broached, unsettling in the extreme. Philosophers and theologians have not settled the questions of the meaning of life and what to do about it, after thousands of years of trying. It would surely be petty to expect evolutionists, particularly operating as they must under the handicap of a lack of divine revelation, to have discovered – or revealed – the true dogma already.

The genesis of this volume came at a session for the History of Science Society meeting in Vancouver (2000) and in a subsequent conference held at the Max Planck Institute for the History of Science, Berlin (2000). The editors express their gratitude to the institute and its staff, particularly Sylvia Knaust and Carola Kuntze; to Institute Director Lorraine Daston; and to the Max Planck Society for the support that made this volume possible.

CHAPTER TWO

Russian Theoretical Biology between Heresy and Orthodoxy

Georgii Shaposhnikov and His Experiments on Plant Lice

Daniel Alexandrov and Elena Aronova

Everyone is heretical, everyone is orthodox.

Umberto Eco, *The Name of the Rose*

Such expressions as "orthodox Darwinian" and "unorthodox theorist" don't usually require definition. For each scientist in a conversation understands clearly what "orthodox Darwinian" means. At the same time, it is obvious that the understanding of "orthodox Darwinian" changes with time and depends on context. Darwinian evolutionary theory became commonplace among biologists, a part of the basic canon of life sciences. The main disputes were about whether Darwinian biology was the best one, whose position was the most orthodox, and whose positions were heretical. Undoubtedly, at certain periods what would have been considered orthodox among American geneticists would have been perceived as heretical among French experts, and vice versa. This chapter deals with the history of Russian evolutionary biology from 1930s to the 1980s. We will attempt to show how in Russian biology the very idea of what being an orthodox evolutionist meant and what being a heretic meant was modified in different contexts.

In our study, we use the life and work of one entomologist as a case study. Georgii Shaposhnikov (1915–1997) was a taxonomist of plant lice (*Aphidoidea*), a group of insects with a peculiar life cycle and biology. His experiments producing rapid speciation in aphids were widely discussed by major players in the field of evolutionary biology, as the experiments seemed to fit well into many theoretical frameworks. At various times, these experiments were used as arguments to support quite different evolutionary mechanisms. Shaposhnikov's research in experimental evolution both gained him popularity among evolutionary biologists of all theoretical denominations and brought him into the field of theoretical biology itself. Shaposhnikov was by no means a famous biologist, but his case nevertheless allows us to review all the major discussions in theoretical biology during his career: from the discussions of the 1940s and 1950s, when Lysenko was in power, to the 1970s and 1980s, when such ideas as systems theory, discontinuities in evolution, and new searches for the morphological laws of evolution came into fashion. In the course of the chapter, we will alternate between description of the general context of theoretical debates and the particular history of Shaposhnikov's research.

The Russian case allows us also to trace the distinct political and scientific contexts of orthodoxy and heterodoxy in Russian biology, for at Lysenko's time, to be an orthodox geneticist was equal to being a political heretic. By comparing the views of Russian biologists with those of their colleagues in different countries, we hope to show that the disposition of *authority of knowledge*, which determines the perception of one or another epistemic position as orthodox or heretical during the period of biology's successful development, turned out to be almost diametrically opposed: scientists with identical views were considered orthodox in one country and heretical in another.[1]

---

[1] Such phenomena are well known in the sociological concept of "scientific fields" introduced by P. Bourdieu, which we use for the purposes of our analysis. See Pierre Bourdieu, "The Specificity of the Scientific Field and the Social Conditions of the Progress of Reason," *Social Science Information* 14 (1975): 19–47; Pierre Bourdieu, "The Peculiar History of Scientific Reason," *Sociological Forum* 6 (1991): 3–26; and for general discussion of literature and culture; Pierre Bourdieu, *The Field of Cultural Production: Essays on Art and Literature* (Cambridge, UK: Polity Press, 1993). Bourdieu defines a scientific field as "a field of forces whose structure is defined by the continuous distribution of the specific capital possessed, at the given moment, by various agents or institutions operative in the field. It is also a field of struggles or a space of competition where agents or institutions who work at valorizing their own capital – by means of strategies of accumulation imposed by the competition and appropriate for determining the preservation or transformation of the structure – confront one another" (Bourdieu, "The Peculiar History," p. 7).

   In this chapter we do not strictly follow Bourdieu's sociological theory in all of its ramifications and implications; we merely find the concept of a scientific field useful for the purpose of representation and analysis of our material. Fritz Ringer used these ideas of Bourdieu's in a similar way in his social intellectual history of French intellectuals. Ringer especially concentrated on the interrelationship of orthodoxy and heterodoxy: "The views expressed in a given setting are so thoroughly interdefined that they can be adequately characterized only in their complementary or oppositional relationships to each other. Indeed, opposed positions within an intellectual field tend to condition each other. The prevailing orthodoxies of a given context help to shape the heterodox reversals they call into being, and of course they determine the

RUSSIAN EVOLUTIONARY BIOLOGY, 1930s–1950s

Theoretical and evolutionary biology flourished in Soviet Russia from the 1920s to the 1940s despite all the trials and tribulations that scientists experienced. In the 1920s, nearly as many theoretical positions were espoused as there were active scientists – genetics and eugenics, embryology, evolution, and ecology were debated and developed in Russia from various theoretical perspectives within a rather diverse institutional landscape.[2] The diversity of publicly expressed intellectual positions diminished under the ideological pressure of several political campaigns in the 1930s, but multifocal debates remained a part of scientific life in biology.

Lamarckism in its various strands was also constantly debated by its supporters and opponents.[3] Among Russian neo-Lamarckians of the 1920s were scientists of different ages, backgrounds, and research interests: from old

---

structure of the field as a whole. At the same time, heterodox ideas may well acquire a certain dominance in their own right" (Fritz Ringer, *Fields of Knowledge: French Academic Culture in Comparative Perspective, 1890–1920* [Cambridge: Cambridge University Press, 1992], p. 4). See also his article and the following discussion: Fritz Ringer, "The Intellectual Field, Intellectual History and the Sociology of Knowledge," *Theory and Society* 19 (1990): 269–94. It is evident that the discussion of any intellectual heresy implies the analysis of the respective intellectual fields in which the intellectual claims under scrutiny were made.

[2] See Mark B. Adams, "Towards a Synthesis: Population Concepts in Russian Evolutionary Thought, 1925–1935," *Journal of the History of Biology* 3 (1970): 107–29; Mark B. Adams, "Science, Ideology and Structure: The Koltsov Institute, 1900–1970," in L. L. Lubrano and S. G. Solomon, eds., *The Social Context of Soviet Science* (Boulder, CO: Westview Press / Folkestone, England: Dawson, 1980): 173–204; Mark B. Adams, "Sergei Chetverikov, the Koltsov Institute, and the Evolutionary Synthesis," in E. Mayr and W. B. Provine, eds., *The Evolutionary Synthesis: Perspectives on the Unification of Biology* (Cambridge, MA: Harvard University Press, 1980): 242–78; M. B. Adams, ed., *The Wellborn Science: Eugenics in Germany, France, Brazil, and Russia* (New York and Oxford: Oxford University Press, 1990); Theodosius Dobzhansky, "The Birth of the Genetic Theory of Evolution in the Soviet Union in the 1920s," in Mayr and Provine, eds., *The Evolutionary Synthesis*, pp. 229–42; Abba E. Gaissinovich, "The Origin of Soviet Genetics and the Struggle with Lamarckism," *Journal of the History of Biology* 3 (1980): 1–51; Abba E. Gaissinovich, *Zarozhdenie i razvitie genetiki* [The origin and development of genetics] (Moskva: Nauka, 1988); Loren R. Graham, "Science and Values: The Eugenics Movement in Germany and Russia in the 1920s," *American Historical Review* 82 (1977): 1135–64; Loren R. Graham, *Science in Russia and the Soviet Union: A Short History* (Cambridge: Cambridge University Press, 1993); Nikolai L. Krementsov, *Stalinist Science* (Princeton, NJ: Princeton University Press, 1997); Douglas R. Weiner, *Models of Nature: Ecology, Conservation and Cultural Revolution in Soviet Russia* (Bloomington: Indiana University Press, 1988).

[3] See Gaissinovitch, "The Origin of Soviet Genetics"; Lev Ia. Bliakher, *The Problem of the Inheritance of Acquired Characters: A History of A Priori and Empirical Methods Used to Find a Solution*, English translation edited by F. B. Churchill (New Delhi: Amerind, 1982); Douglas R. Weiner, "The Roots of 'Michurinizm': Transformist Biology and Acclimatization as Current in the Russian Life Science," *Annals of Science* 42 (1985): 243–60.

traditional professors, for whom it was their nineteenth-century legacy, to young biologists of socialist inclinations, for whom it was a part of modern experimental biology and modern evolutionary theory. Some young Russian Lamarckians during the 1920s involved themselves in genetic laboratory research and were quickly converted to chromosomal genetics, but others remained faithful and carried the debate into the more general realm of evolutionary theory.

In the early 1930s, Trofim Lysenko and his supporters promulgated their views of inheritance and their criticism of genetics and soon became the major proponents of crude Lamarckism. Indeed, that it was possible for Lysenko to join in major academic debates was very much due to the diversity of opinions still present in the scientific field of biology during this period. Lysenko's story is a textbook example of Bourdieu's theory of symbolic capital – his rise at the beginning was based not on direct patronage from the highest Soviet authorities but on his skillful conversion of scientific into political capital and back again.[4]

The role of adaptive modifications in evolution, the inheritance of acquired characteristics, and the interpretation of various experiments adduced by Lamarckians to back up their claims were subjects of lively discussion during the 1920s and 1930s. The lack of conclusive evidence for Lamarckian claims about the possibility of the hereditary fixation of adaptations, as well as difficulties in the interpretation of some of their experiments, justified the dominant opinion that modifications were not transmitted to progeny and therefore do not play a significant role in evolution.[5] In the face of growing Lysenkoism, a number of Russian biologists raised anew the question of the role of adaptive modifications in evolution.[6]

---

[4] See various treatments of Lysenko's career: David Joravsky, *The Lysenko Affair* (Cambridge, MA: Harvard University Press, 1970); Nils Roll-Hansen, "A New Perspective on Lysenko?," *Annals of Science* 42 (1985): 261–78; Valery Soyfer, *Vlast' i nauka: Istoriia razgroma genetiki v SSSR* [Power and science: The history of genetics' defeat in USSR] (Moskva: Lazur, 1993); Valery Soyfer, *Lysenko and the Tragedy of Soviet Science* (New Brunswick, NJ: Rutgers University Press, 1994).

[5] As Georgii Gause wrote, interest in the studies of modifications declined when it became clear that adaptive modifications are not hereditary changes: see Georgii F. Gause, "Ekologiia i nekotorye problemy proiskhozhdeniia vidov [Ecology and some problems of the origin of species]" (1941), first published in Y. M. Gall, ed., *Ekologiia i evolutsionnaia teoriia* [Ecology and the evolutionary theory] (Leningrad: Nauka, 1983), pp. 5–105.

[6] Efim Lukin, a zoologist from Kharkov, considered in his theoretical paper "On the Causes of Substitution of Modifications by Mutations in the Process of Organic Selection from the Viewpoint of the Theory of Natural Selection" the parallelism between phenotypic and genotypic variability, and he arrived at the following conclusions: (1) Organisms frequently respond to

Ivan Schmalhausen, in his theory of stabilizing selection, dealt most comprehensively with these problems.[7] Schmalhausen attached great importance to the role of phenotypic variability in evolution. He believed that adaptive

environmental changes by adaptive phenotypic modifications. (2) Similar adaptive characters may be genotypically fixed in races normally living in the corresponding environments. (3) It has been proved that conversion of modifications into mutations is not possible. (4) Hence, modifications can only be substituted by coincident mutations if the latter are associated with some advantage in the process of natural selection. Efim I. Lukin, "On the Causes of Substitution of Modifications by Mutations in the Process of Organic Selection from the Viewpoint of the Theory of Natural Selection," *Uchenye zapiski Khar'kovskogo Universiteta* 6 (1936): 199–209 [in Ukrainian]. This work was reviewed by Georgii F. Gause in his "Problems of Evolution," *Transactions of the Connecticut Academy of Arts and Sciences* 37 (1947): 17–68. See also Efim I. Lukin, "Adaptivnye nenasledstvennye izmeneniia organizmov i ikh sud'ba v evolutsii [The adaptive nonhereditary modifications and their destiny in evolution]," *Zhurnal Obshchei Biologii* 3 (1942): 235–61.

   The same problems were discussed at that time by Valentin Kirpichnikov, a young geneticist from the Koltsov Institute in Moscow, who proposed the theory of "coincident selection." See Valentin S. Kirpichnikov, "Rol' nenasledstvennoi izmenchivosti v protsesse estestvennogo otbora (gipoteza o kosvennom otbore) [The role of nonhereditary variability in the process of the natural selection (hypothesis of the indirect selection)]," *Biologicheskii Zhurnal* 4 (1935): 775–800; Valentin S. Kirpichnikov, "Znachenie prisposobitel'nykh modifikatsyi v evolutsii [The meaning of the adaptive modifications in evolution]," *Zhurnal Obshchei Biologii* 1 (1940): 121–52. He argued that in the presence of adaptive modification, natural selection will improve this modification. He considered fixation of adaptive modifications as the indirect result of direct selection. Similar views were expressed by Georgii Gause, an experimental biologist well known for his work on experimental ecology. (See Sharon Kingsland, *Modeling Nature: Episodes in the History of Population Ecology* (Chicago: University of Chicago Press, 1985). In his earlier works, Gause argued that new adaptations arise as adaptive modifications, which are then fixed by means of natural selection. (See, for instance, Gause's review of Schmalhausen's monograph *Organizm kak tseloe v individual'nom i istoricheskom razvitii* in the *Quarterly Review of Biology* 14 [1939]: 65–7.) Later, he discussed only the substitution of modifications by mutations through natural selection (see Gause, "Ekologiia i nekotorye problemy proiskhozhdeniia vidov").

[7] Ivan I. Schmalhausen, *Organizm kak tseloe v individual'nom i istoricheskom razvitii* [Organism as a whole in the individual and historical development](Moskva–Leningrad: Izdatel'stvo AN SSSR, 1938); Ivan I. Schmalhausen, *Puti i zakonomernosti evolutsionnogo protsessa* [The ways and the laws of the evolutionary process](Moskva–Leningrad: Izdatel'stvo AN SSSR, 1940); Ivan I. Schmalhausen, *Faktory evolutsii (teoriia stabiliziruiushchego otbora* [Factors of evolution (the theory of stabilizing selection)] (Moskva–Leningrad: Izdatel'stvo AN SSSR, 1946). The English edition of *Faktory evolutsii* is Ivan I. Schmalhausen, *Factors of Evolution: The Theory of Stabilizing Selection*, trans. I. Dordick, ed. T. Dobzhansky (Philadelphia: Blakiston, 1949). On Schmalhausen, see Mark B. Adams, "Severtsov and Schmalhausen: Russian Morphology and the Evolutionary Synthesis," in E. Mayr and W. B. Provine, eds., *The Evolutionary Synthesis: Perspectives on the Unification of Biology*. (Cambridge, MA: Harvard University Press, 1980), pp. 197–204; Dobzhansky, "The Birth of the Genetic Theory," note 2; Yakov M. Gall, "I. I. Shmal'gauzen i problema faktorov evolutsii" [I. I. Schmalhausen and the problem of factors of the evolution]," in S. R. Mikulinskii et al., eds., *Istoriko-biologicheskie issledovaniia*, vol. 8. (Moskva: Nauka, 1980), pp. 106–23.

modifications mark the path of the evolution of species. He argued that adaptive modifications could be substituted for by mutation, though not precisely: selection evaluates the advantages of genocopies according to their general adjustment, but not according to their adequacy to morphoses. Some Russian followers of Schmalhausen's approach linked these new concepts with the old Baldwin–Morgan notion of organic selection. As one of them wrote, "the terms organic, stabilizing, coincident selection are practically synonymous. . . . The theoretical studies by Lukin, Schmalhausen, Kirpichnikov reveal so very much in common with those of Baldwin, Lloyd Morgan and Osborn. . . ."[8] By contrast, Theodosius Dobzhansky, who promoted Russian researchers in the West whenever he could, gave a modern reading to Schmalhausen's approach, placing it along with Waddington's theory: "As pointed out by . . . Schmalhausen (1949), [and] Waddington (1957, 1960) . . . development may be canalized to yield a fixed outcome or may be plastic to produce varying phenotypes adaptive in different environments. The term organic selection has been coined to describe the parallelism between racial genotypic and environmental phenotypic variability."[9]

Schmalhausen thus combined the epigenetic approach with the theory of natural selection, thereby stripping Lamarckians of their "trump card." Moreover, Schmalhausen made the problem of the inheritance of acquired characters senseless, since according to his approach individual adaptability is considered the expression of hereditary qualities. Thus, instead of the traditional distinction between "hereditary" and "acquired" characters, all characters in the framework of Schmalhausen's approach are considered to be stable under the external influence or dependent on it.

All of these theories explained the role of phenotypic adaptation in evolution within Darwinian frameworks. Their simultaneous appearance in Russia seemed to be a response to attempts to treat the phenomenon of adaptive modification as proof of Lamarckian mechanisms of evolution. According the concept of "scientific field," these theories can be viewed as strategies – counteracting the attacks on genetic theory of natural selection and its proponents, and showing that the existing scientific field of evolutionary and theoretical biology in Russia at that time still allowed for different strategies predicated on epistemic moves rather than the sheer use of political capital.[10]

[8] Gause, "Problems of Evolution," p. 22.
[9] Theodosius Dobzhansky, *Genetics of the Evolutionary Process* (New York and London: Columbia University Press, 1970), p. 303.
[10] Another illustration of this is the fact that in the first years after World War II, major works by Ernst Mayr and George Gaylord Simpson were published in Russian translation, which indeed

In the 1940s, Ivan Schmalhausen occupied the dominant position in the field of evolutionary and theoretical biology in Russia. He was widely read and published, and he occupied key positions in academic structures: head of the Department of Darwinism at Moscow State University, which was the central university in the highly centralized Soviet "empire of knowledge";[11] director of Severtsov's Institute of Evolutionary Morphology in the Academy of Sciences, the only institute concerned directly with evolutionary biology; and editor-in-chief of *Zhurnal Obshchei Biologii* [The Journal of General Biology], a journal devoted mainly to evolution, embryology, and genetics that he founded in 1940. His epistemic strategy was also highly efficient, as it seemed to provide a "third way" between neo-Darwinism and Lamarckism, apparently resolving the extant controversy.[12]

The infamous VASKhNIL session in August 1948, which made Lysenko and his supporters virtually omnipotent, drastically changed the scene. The session was sanctioned by the highest Soviet authorities, and Lysenko's speech at the session was authorized by Joseph Stalin himself.[13] It effectively silenced debate and crushed the intellectual field of evolutionary and theoretical biology. The field as such virtually disappeared, as scientists were no longer able to express publicly intellectual positions on evolutionary issues that were different from those prescribed by political authorities. Moreover, the political assault on the scientific community drastically reshaped all positions (though this was mainly hidden from the eyes of the public) – all scientists seemed to be divided into open supporters or clandestine opponents of Lysenko.[14]

---

was a strategy of their translators within the existing intellectual field: Ernst Mayr, *Sistematika i proiskhozhdenie vidov* (Moscow: Gosudarstvennoe Izdatel'stvo Inostrannoi Literatury, 1947) (the translation of Ernst Mayr, *Systematics and the Origin of Species* [New York: Columbia University Press, 1942]); George G. Simpson, *Tempy i formy evoliutsii* (Moscow: Gosudarstvennoe Izdatel'stvo Inostrannoi Literatury, 1948) (the translation of George G. Simpson, *Tempo and Mode in Evolution* [New York: Columbia University Press, 1944]).

[11] The main student textbook in Darwinism and evolutionary theory at that time was by I. Schmalhausen, *Problemy darvinizma. Posobie dlia VUZ'ov* [Problems of Darwinism: Textbook for institutes of higher education] (Moskva: Sovetskaia Nauka, 1946).

[12] Even Russian critics of Darwinism were impressed; the most persistent critic of selectionist theories among biologists in Soviet Russia, Alexander Liubishchev, called the theory of stabilizing selection "the most brilliant defense of the most hopeless affair." A. A. Liubishchev to E. S. Smirnov. October 29, 1945 (ARAN, f. 2079).

[13] See, on Stalin's personal involvement: Kirill Rossianov, "Stalin as Lysenko's Editor: Reshaping Political Discourse in Soviet Science," *Configurations* 1 (1993): 439–56; Kirill Rossianov, "Editing Nature: Joseph Stalin and the 'New Soviet Biology'," *Isis* 84 (1993): 728–45.

[14] For example, many scientists who initially disagreed with chromosomal genetics, or at least had doubts about it in the early 1930s, now in the face of Lysenko's political success were united with geneticists, despite their conceptual differences. (For a brief treatment of these

Studies of adaptive modifications and epigenetic evolution were also abandoned. Schmalhausen was stripped of all his administrative power and found refuge as a researcher in the Zoological Institute of the Academy of Sciences (the institute itself was in Leningrad [St. Petersburg], but its administration allowed Schmalhausen to continue working in Moscow).

However, it was for only four years that scientists kept silent on theoretical issues amid praise of Lysenko and his "scientific achievements." The possibility of open opposition to Lysenko arose when Lysenko suggested his "theory of speciation" on the basis of his concept of heredity. Lysenko proposed that under the influence of external conditions, "corpuscles" of a new species are self-conceived in the "body of an old species," and, according Lysenko, this process happens by leaps – old species constantly generating ("giving birth to," as Lysenko would say) the organisms of a new species.[15] Lysenko maintained that this explained the persistence of weeds (species of cultivated plants produce various species of weeds), and he even went so far as to claim that various birds produced cuckoos as their offspring. At first, this "new theory of biological species" offered by the official leader of Soviet biology engendered a myriad of articles providing evidential support for the theory. But Lysenko's new theory and its forged supporting evidence were soon used for a major assault on Lysenko by many biologists.[16] Lysenko proposed his "species theory" two years after the VASKhNIL session, and this new claim evidently had no special sanction from the Communist Party leadership, thus permitting debate.[17] Beginning in 1952, several journals not editorially controlled by Lysenko and his supporters were flooded with articles dealing with the theory of speciation. The first articles to appear were directed against Lysenko and his ideas, but discussion soon radiated outward, as authors discussed almost all aspects of species problems – from gradual geographic speciation as a main mode of evolution, to sympatric "ecological" speciation, to genetic speciation, according to which reproductive isolation of a "new species form" (as they

changes, see Daniel Alexandrov, "Historical Anthropology of Science in Russia," *Russian Studies in History* 34 (1995): 62–91.

[15] Trofim D. Lysenko, "Novoe v nauke o biologicheskom vide [The new in the science of biological species]," *Agrobiologiia* 6 (1950): 15–25.

[16] See the details in Soyfer, *Vlast' i nauka*; Soyfer, *Lysenko and the Tragedy*; Vladimir Ia. Aleksandrov, *Trudnye gody sovetskoi biologii* [The hard years of Soviet biology] (St. Petersburg: Nauka, 1992).

[17] On this interview with the rector of Leningrad University in the 1950s, see Eduard I. Kolchinskii, "Interv'iu s akademikom A. D. Aleksandrovym [Interview with A. D. Alexandrov, member of the Academy]," in M. G. Iaroshevskii, ed., *Repressirovannaia nauka*, vol. 2 (St. Petersburg: Nauka, 1994), pp. 169–76.

called it) precedes the emergence of a new species as a set of populations.[18] The quick shifting of the debate from pure "defense of science" to productive scientific discussion opened a new phase in Russian theoretical biology, making the topic of species and speciation a hot one at that time.

At the same time, several institutions controlled by Lysenko's opponents provided public forums for the discussion of theoretical issues, including the Zoological Institute of the Academy of Sciences, located in St. Petersburg (Leningrad).[19] The seminars and conferences, which discussed subjects dissenting from Lysenkoite orthodoxy, attracted much attention. The very existence of forcefully expressed opposing viewpoints made the field of evolutionary biology vibrant and attractive to scientists. Paradoxically, our interviewees who were participants in and witnesses to these meetings recollected that time as "intellectually exciting," explaining in this way the attraction of evolutionary issues among many scientists at that time, especially among young people.[20]

[18] One has to note that geneticists were almost completely absent among the authors engaged in the debates of the 1950s (they for the most part kept a low profile in these years); the discussion was carried on by botanists, zoologists, and paleontologists. Leading botanists and zoologists discussed in detail the geographic mode of speciation in plants (e.g., A. I. Tolmachev, "O nekotorykh voprosakh teorii vidoobrazovaniia [On some ussues of the theory of speciation]," *Botanicheskii Zhurnal* 38 [1953]: 530–55) or speciation through hybridization (e.g., B. K. Shishkin, "Vystuplenie na diskussii o vide [Speech delivered on the discussion of species]," *Vestnik LGU* 10, no. 4 [1954]: 43–5), or engaged in discussion of Mayr's ideas on semi-species (e.g., G. P. Dement'ev, "Zamechaniia o vide i nekotorykh storonakh vidoobrazovaniia v zoologii [Some remarks on species and on some aspects of speciation in zoology]," *Zoologicheskii Zhurnal* 33 (1954): 525–37). The specific taxonomic groups they worked on often influenced their choice of preferred mode of speciation and their criticism of species concepts. The biological species concept was widely accepted, but many voiced concern over reproductive isolation as a main species criterion; those who worked on agamous or parthenogenetic organisms were particularly worried. (See the review of the discussion in Kiril M. Zavadskii, *Uchenie o vide* [The concept of species] [Leningrad: Izdatel'stvo LGU, 1961]; Kiril M. Zavadskii, *Vid i vidoobrazovanie* [Species and speciation] [Leningrad: Nauka, 1968]).

[19] In Leningrad, the other institutions that provided relatively free forums for publication, seminars, conferences, and public speeches during the Lysenko period were several institutes of the Academy of Sciences, the Botanical Society, Leningrad University, and the Leningrad Society of Naturalists attached to the university. In Moscow, where academic as well as political seats of power were concentrated, the only open forum was the Moscow Society of Naturalists, which was a voluntary association, independent of the Academy of Sciences. By virtue of its democratic organization and relative independence of authority, this society became an umbrella organization for many scientific dissenters from the 1950s to the 1980s. See the excellent discussion of its activity in the mid-1950s in Douglas R. Weiner, *A Little Corner of Freedom* (Berkeley: University of California Press, 1999).

[20] Interview with Vladimir I. Kuznetsov of the Zoological Institute, St. Petersburg, October 3, 2000.

Thus, that the scientific field in a very Bourdieuian sense reemerged successfully in the years from 1952 to 1956 was due to the vivid discussion stirred up by Lysenko's theory of speciation. Species and speciation became the major issue defining positions in the area of evolutionary biology. The field itself was polarized – it unfolded from a virtually one-dimensional state, with no public dissent from Lysenkoism, to something more like a rhombus, with the major axis still being pro- or anti-Lysenkoism. Scientists may have disagreed with one another on modes of speciation, or any other big issue in evolutionary theory, but most of them were united in their opposition to Lysenko. In the mid-1950s, this polarization temporarily brought together people who had held opposing views in the 1930s: staunch anti-Darwinians such as the zoologist Alexander Liubishchev, and Darwinians such as Ivan Schmalhausen.[21]

This was the general context in which Georgii Shaposhnikov began his experimental studies of speciation in aphids.

GEORGII SHAPOSHNIKOV'S EXPERIMENTS WITH PLANT LICE AND THE SHIFT TO THEORETICAL BIOLOGY

Georgii Khristoforovich Shaposhnikov, son of a well-known naturalist and the organizer of the Caucasian State Nature Reserve (*zapovednik*), was born in Maikop on February 18, 1915. Georgii's childhood was spent in the milieu of prominent Russian zoologists. The most important figure in his life and career was the leading Russian aphidologist Alexander Mordvilko.[22] Shaposhnikov graduated in 1938 from the Leningrad Agricultural University as an agronomist specializing in a plant protection. During the period of the Stalinist reprisals, his father was arrested and in 1938 executed. As a

---

[21] Daniil A. Alexandrov, "Teoreticheskaia biologiia: edinstvo dvizheniia i raznost' idei [Theoretical biology: Unity of movement and diversity of ideas]," *Priroda* 9 (1989): 121–3.

[22] As Shaposhnikov recalled: "I regard myself as his [Mordvilko's] disciple, as I spent a lot of time studying his collections, books and papers. The young Mordvilko worked at Warsaw University, where he made many observations in the Botanical Garden and in nature.... Then he worked at the Zoological Museum (later Institute) in St. Petersburg–Leningrad. In 1925 Mordvilko spent some days in Maikop (a town in the North-West Caucasus) in my father's house. My father as director of Caucasus Nature Reserve was very busy at that time. I was ten years old and acted as Mordvilko's guide in the environs of Maikop. Both entomologists, Mordvilko and my father, wanted me to be an aphidologist, but I preferred beetles, butterflies and dragonflies. Nevertheless both gentlemen forecast my fate correctly...." Georgii Ch. Shaposhnikov and Andrei V. Stekolshchikov, "Progress of Aphidology in the Twentieth Century," in J. M. Nieto Nafria and A. F. G. Dixon, eds., *Aphids in Natural and Managed Ecosystems* (Leon: Universidad de Leon, 1998), pp. 27–35.

result, in spite of Mordvilko's desire to have him in the Zoological Institute in Leningrad as his graduate student, Shaposhnikov failed to get a position in Leningrad and became instead quarantine inspector in the Altay region of southern Siberia. Mordvilko's support helped him to return to Leningrad, and in January 1941 he was appointed to the staff of the Zoological Institute. However, his work at the Zoological Institute began only in 1946, after his service in the army and then in the Military Medical Academy (Voenno-Meditsinskaia Academiia).

The Zoological Institute of the Academy of Sciences was fundamentally an old, established natural history museum with enormous collections dating back to the eighteenth century, which made it the central Russian research establishment in taxonomy.[23] It was spared the move in 1934 from Leningrad to Moscow, when most of the institutes and laboratories of the Academy of Sciences, which previously had been situated in Leningrad (the former Russian imperial capital of St. Petersburg), were moved to the Soviet capital city.[24] The institute enjoyed relative independence in the Academy under the leadership of Evgenii Pavlovskii – a well-known parasitologist, a shrewd politician, and a very powerful figure as a full member of the Academy, a two-star general of military medical service, and the head of the Department of Biology and Parasitology in the Military Medical Academy. Pavlovskii calculated every move, advancing his institute by making compromises without scruples: he fired one geneticist (Valentin Kirpichnikov, mentioned earlier) to prove that he was ruthless with Lysenko's enemies, published an outrageous research report of experimental proof of Lysenkoan inheritance, and very soon hired several prominent scientists (Ivan Schmalhausen among them) who had lost their jobs in other institutes.[25] Pavlovskii himself never opposed Lysenko, but he was editor-in-chief of *Zoologicheskii Zhurnal* and *Entomologicheskoe Obozrenie*, which under his editorship published very interesting papers on evolution that rather freely dissented from contemporary orthodoxy.

When Shaposhnikov returned in 1946 to the Zoological Institute, he inherited Mordvilko's "aphid room" and his rich collection of aphids in the Department of Insect Taxonomy. Shaposhnikov acted energetically in his new field and got in touch with other aphidologists, who viewed Shaposhnikov

---

[23] See O. A. Skarlato, ed., *Zoologicheskii institut. 150 let* [The Zoological Institute: 150 years] (Leningrad: ZIN AN SSSR, 1982).

[24] As a result of this relocation, the Academy of Sciences was supposed to become, and in fact became, part of government control apparatuses, and of course as a result government control over the Academy itself was tightened. (On the system of Stalinist science, see Krementsov, *Stalinist Science*).

[25] See the eyewitness account of Pavlovskii's dealings in Aleksandrov, *Trudnye gody*.

as Mordvilko's successor and indeed helped him generously.[26] By the beginning of the 1950s, Shaposhnikov had published a dozen papers concerning the taxonomy and various peculiarities of the lifecycles and biology of aphids, becoming an expert in his group. In his first papers, he cited Lysenko's works (though without emphasizing these references) and connected his research primarily with Pavlovskii's parasitology, thus paying homage to his academic patron. The most noteworthy among these earlier publications is a general review devoted to the evolution of aphids in relation to their host plants, in which Shaposhnikov argued that, like many parasitic groups, aphids follow in general the evolution of their hosts – in this particular case, the evolution of plants: from conifers to leaf-bearing trees, from trees to shrubs, from trees and shrubs to herbaceous plants. In particular, he came to the conclusion that "the main cause of such a direction of evolution in plants and aphids is the same adaptation to life under conditions of a drier climate."[27]

In the course of this research, studying ontogenetic and phylogenetic trends in aphids, Shaposhnikov did some experimental work testing the ability of different aphid species to live on new food plants. In the mid-1950s, he decided to carry out a series of experiments on the adaptation of aphids to new environments. Shaposhnikov was strongly attracted to experimental studies of evolution, especially because he saw in the burgeoning intellectual field of evolutionary biology an opportunity to make advances without being forced to subscribe to extreme positions. In 1957, at the time when Lysenko was in full institutional power, Shaposhnikov noted a positive moment – "the possibility for truly scientific debates and the organization of experiments."[28] To experiments he turned his efforts.

Shaposhnikov began by selecting species of plant lice and their host plants for experiments, which took him almost three years. The main experiment began in 1957 and ended in 1959, and the procedure was as follows:

Offspring of one female aphid (*Dysaphis anthriski majkopica*) (the fundatrice) were divided into three groups. The first group was nurtured on its

---

[26] The community of aphidologists was very small and dispersed in different regions of the USSR. As Shaposhnikov wrote in the late 1940s: "Now we are eight in number: in pairs in the North, in Ukraine, in Central Asia and in the Caucasus." Shaposhnikov to V. A. Mamontova, March 25, 1948, Shaposhnikov papers, Zoological Institute, St. Petersburg.

[27] Georgii Ch. Shaposhnikov, "Evoliutsiia nekotorykh grupp tlei v sviazi s evoliutsiei rozotsvet-nykh [Evolution of some aphid groups in relation to evolution of Rosadeae]," in *Tret'i ezhe-godnye chteniia pamiati N. A. Kholodkovskogo* (Moscow/Leningrad: Akademia Nauk SSSR, 1951), pp. 28–60, at p. 58.

[28] Shaposhnikov to Evgenii S. Smirnov, February 13, 1957, Smirnov papers, Moscow Archive of the Russian Academy of Sciences, f. 2079.

natural host plant (*Anthriscus nemorosa*) for fifty generations without cross-breeding (reproducing parthenogenetically), as a control line. The second group was transferred onto an absolutely unsuitable plant (*Chaerophyllum maculatum*), where all of them died. The third group was transferred onto a poorly suitable plant (*Ch. bulbosum*), where some of them survived and produced offspring (parthenogenetically). Later, some individuals of each surviving generation were transferred back to the original host plant, and some to the absolutely unsuitable plant. It turned out that after eight generations on the poorly suitable host, the aphids had become unable to live on their natural host but had acquired the ability to live on the new, previously absolutely unsuitable host plant. This new form of aphid was able to survive on the absolutely unsuitable plant and continued to reproduce parthenogenetically through the forty-seventh generation, when the experiment was terminated. The new form was different from the parent species, both morphologically and ecologically; moreover, it was unable to interbreed (sexually) with the parent species and thus seemed to be a new species.[29]

Let us see how Shaposhnikov presented these experiments. His rhetoric was "neutral," in accordance with the academic style of taxonomists from the Zoological Institute – he avoided direct alliances with any grand theory of his time and even cited Lysenko in his first articles, but only in passing as one of the authors who had written on the subject.[30] With respect to theoretical interpretation, he claimed that his results showed how "preexisting potentialities" are revealed under intense natural selection and wrote of the importance of "genetic plasticity" in providing the material on which natural selection can act – in discussing the nature of such plasticity, he referred to Lysenko's notion of "shattered heredity." He believed that only clones with the

---

[29] Georgii Ch. Shaposhnikov, "Stanovlenie smeny khoziaev i diapausy u tlei (Aphididae) v protsesse prisposobleniia k godichnym tsiklam ikh kormovykh rastenii [The initiation and evolution of the change of hosts and the diapause in plant lice (Aphididae) in the course of adaptation to the annual cycles of their host plants]," *Entomologicheskoe Obozrenie* 38 (1959): 483–504; Georgii Ch. Shaposhnikov, "Spetsifichnost' i vozniknovenie adaptatsii k novym khoziaevam u tli v protsesse estestvennogo otbora [Specificity and the appearance of adaptations to new hosts in aphids during the process of natural selection]," *Entomologicheskoe Obozrenie* 40 (1961): 739–62; Georgii Ch. Shaposhnikov, "Morfologicheskaia divergentsiia i konvergentsiia v eksperimente s tliami," [Morphological divergence and convergence in an experiment with aphids]," *Entomologicheskoe Obozrenie* 44 (1965): 3–25; Georgii Ch. Shaposhnikov, "Vozniknovenie i utrata reproduktivnoi izoliatsii i kriterii vida [The rise and breakdown of reproductive isolation and the species criterion]," *Entomologicheskoe Obozrenie* 45 (1966): 3–35.

[30] Shaposhnikov, "Stanovlenie smeny khoziaev"; Shaposhnikov, "Spetsifichnost' i vozniknovenie adaptatsii."

"largest plasticity" survived through the process of intense natural selection. But this theoretical part of his research was virtually lost in the abundance of morphological and taxonomic material.

Shaposhnikov presented himself in these papers as a zoologist and entomologist concerned with detail: he gave long naturalistic descriptions of morphology, provided extensive keys to species of plant lice, detailed very minute morphological changes that occurred in the experiments, and so on. The key issue was the definition of the taxonomic status of the new form. Since it did not interbreed with the parental form, Shaposhnikov from the beginning discarded the possibility of its being just a modification. In subsequent publications, he raised the stakes. First he considered the new form to be a new *morpho*[31]; then it was a new intraspecific form (probably a subspecies);[32] and finally it was "a new form of species rank or, at least, very close to species.[33] He argued in these publications on the basis of morphological differences, and despite his belief in the possibility of truly scientific debates in these years, he stuck to good old morphological taxonomy.

This presentation drastically changed in the publication that appeared in 1966 – the first article written and submitted to the journal after Lysenko was officially dethroned in the fall of 1964, soon after Khrushchev's downfall.[34] The paper was published in the same *Entomologicheskoe Obozrenie* (Entomological Review) as all his previous articles, but now Shaposhnikov abandoned the style of natural history. Instead of morphology, Shaposhnikov focused on reproductive isolation – the paper's title was very general and pointed directly to the main theoretical problem to be discussed on the basis of his experiments: "the emergence and disappearance of reproductive isolation and species criteria." It seems that as soon as it was permissible, Shaposhnikov opted for a highly theoretical level of discussion. Most characteristically, he even abandoned (as if for brevity) the use of Latin names in some parts of his article, introducing instead conventional signs (species N, M, C), which underscored his abandonment of naturalistic descriptions and gave his paper rhetorically even more of an analytical flavor. Shaposhnikov also dramatically expanded his repertoire of references – he cited not only major figures such as Th. Dobzhansky, J. Huxley and E. Mayr, but also K. Mather, G. M. Lerner, J. M. Thoday, and many other population geneticists. The question of whether he had actually read all these articles by 1965 – he apparently had not – is almost

[31] Shaposhnikov, "Stanovlenie smeny khoziaev."
[32] Shaposhnikov, "Spetsifichnost' i vozniknovenie adaptatsii."
[33] Shaposhnikov, "Morfologicheskaia divergentsiia."
[34] Shaposhnikov, "Vozniknovenie i utrata."

irrelevant. He clearly was well oriented in the literature, wanted to demonstrate that, and – what is most important – used this literature to position himself on the cutting edge of evolutionary research.[35]

After producing this significant article, Shaposhnikov did not publish a word on theoretical issues for eight years. His main production during these years was in the field of aphid taxonomy. Beginning in 1974, he published a series of articles in *Zhurnal Obshchei Biologii* (Journal of General Biology); almost every year the journal published one substantial article by Shaposhnikov, all of which were concerned with major issues then being debated in Russian theoretical biology. These publications again are very different from what he had published previously, and his experiments were almost never mentioned. His now addressed such themes as systems theory in relation to biology; the hierarchical organization of living systems; the systemic approach to the evolution of populations, species, and ecosystems; and the problem of directed evolution and teleonomy.[36]

In 1978, Shaposhnikov returned to the analysis of his experiments.[37] In a lengthy review entitled "Evolution and the Dynamics of Clones, Populations and Species," he considered three types of adaptive transformations: regulatory, phyletic, and quantum. His experiments with plant lice were considered quantum transformations. In this paper, Shaposhnikov followed George Gaylord Simpson in distinguishing different forms of selection acting on different stages of evolution,[38] but he combined Simpson's approach with a lingo

[35] Toward the end of the 1940s, Shaposhnikov began compiling a card index of references that shows his technique of writing and referencing (Shaposhnikov papers). Many articles he referred to were known to him from abstracts provided in *Referativnyi Zhurnal Biologiia* (a Russian analogue of *Biological Abstracts*) or from various monographs he had read. He read only the most interesting articles and made notes to that effect on the cards. Shaposhnikov's card index is a good illustration of how scientists use citation to position themselves in the scientific field. For example, the cards with abstracts of Lysenko's publications were just for "internal use" – Shaposhnikov never cited them.

[36] Georgii Ch. Shaposhnikov, "Vzaimozavisimost' zhivykh sistem i estestvennyi otbor [Interrelations of living systems and natural selection]," *Zhurnal Obshchei Biologii* 35 (1974): 196–208; Georgii Ch. Shaposhnikov, "Zhivye sistemy s maloi stepen'iu tselostnosti [Living systems with a small degree of integrity]," *Zhurnal Obshchei Biologii* 36 (1975): 323–35; Georgii Ch. Shaposhnikov, "Ierarkhiia zhivykh sistem [The hierarchy of living systems]," *Zhurnal Obshchei Biologii* 37 (1976): 493–505; Georgii Ch. Shaposhnikov, "Napravlennost' evoliutsii [The direction of evolution]," *Zhurnal Obshchei Biologii* 38 (1977): 649–55; Georgii Ch. Shaposhnikov, "Dinamika klonov, populiatsii i vidov i evoliutsiia [Dynamics of clones, populations and species, and evoluion]," *Zhurnal Obshchei Biologii* 39 (1978): 15–33.

[37] Shaposhnikov, "Dinamika klonov."

[38] Simpson, *Tempo and Mode in Evolution* and its Russian translation (see note 10); George G. Simpson, *The Major Features of Evolution* (New York: Columbia University Press, 1961).

drawn from systems theory. Shaposhnikov argued that the rapid evolution in his experiments could be considered as quantum evolutionary change, and that a special form of natural selection was at work at the key moment of transformation. He called this selection, which resulted in the destabilization of the "old system" and the formation of a new one, "selection for revealing potencies." After a few generations, he argued, this form of selection had been replaced by the "organizing" selection that fixed the new adaptive state.[39]

Shaposhnikov's interest in systems theory was shared by many biologist of his time in Russia. Indeed, systems theory was a discourse that allowed scientists to speak of "living organization," "integrity," and "directed processes" without feeling that they had to resort to the vitalism or metaphysics that were so abhorred by positivist-minded scientists. This flexibility had made systems theory appealing and fashionable worldwide, but in Soviet Russia it probably had an even stronger appeal. Systems theory in Russia in the 1960s and 1970s provided a legitimate scientific discourse that could be used to discuss problems in a way that would otherwise be considered dangerously idealistic by the Soviet ideological watchdogs. There is no evidence that Shaposhnikov ever reflected on his interest in systems theory, and such reflection was probably not his style. Systems theory was a hot topic at that moment, and he evidently always followed modern trends, being able and even eager to learn new languages that could then be used for the description and understanding of evolutionary phenomena.

Shaposhnikov's theoretical publications during the 1970s were the result of the intensive scientific communication that he was engaged in at the time. He traveled a great deal, attended many conferences, participated in many seminars and summer schools on evolutionary and theoretical biology, and made the acquaintance of all of the active theoretical biologists of the time. By the late 1970s, he was fully immersed in the life of this community, becoming part of a "theoretical crowd." In order better to understand Shaposhnikov's shift to theoretical biology, we need first to review this scientific field in Russia for the period of the 1960s through the 1980s.

THEORETICAL BIOLOGY IN RUSSIA, 1960s–1980s

The process of developing the scientific field of evolutionary and theoretical biology in Russia from the mid-1950s to the 1970s was very exciting for the participants, who felt they were part of something really important. Starting

---

[39] Shaposhnikov, "Dinamika klonov."

with the first discussions of species and speciation, which had the importance and excitement of civil activism – a war waged against Lysenko and therefore against the oppression of science – all scientific dissent in the well-controlled orthodoxy of Soviet Russia was regarded as bordering on civil activism and thus acquired unusual symbolic status for the dissenters. To be a proponent of synthetic theory of evolution in the 1960s was truly exciting, in the same way as it became exciting to be critical of orthodox Darwinism a decade later.

Most of the centers of theoretical activity remained on the periphery, both institutionally and geographically: they were either far away from Moscow centers of power or independent of the Academy of Sciences. New centers provided refuge for theoretical heterodoxy in Estonia (Tartu University and the Estonian Society of Naturalists); in Novosibirsk, with its *Akademgordok* (Academytown) built in the 1960s; and in Pushchino and other new research centers, built in the 1960s in the Moscow region but at some distance from Moscow. These new research centers were created outside of Moscow on purpose – the leaders of the Academy of Sciences recognized the danger of solidified scientific structures and tried to diversify the institutional landscape of Soviet science in order to boost its productivity.[40] Summer schools and conferences hosted in these places became favorite pastimes for many scientists, irrespective of their age.

Institutional peripherality also worked well for heretics. The Moscow Society of Naturalists, with its permanent seminars on evolution and classification, was still an umbrella organization for the Moscow heretics. Another peripheral institution used by theoretical biologists was the Institute for the History of Science and Technology of the Academy of Sciences, with its Moscow and Leningrad branches. It was indeed marginal from the point of view of big-science scientists and ideological bosses, so those who were not fit for either big science or orthodox philosophy in the Academy of Sciences were exiled to this institute.[41] History of science in Russia, despite the existence of a research institute, never became an integrated discipline with a common language and research agenda across disciplinary histories – historians of biology always regarded biologists, and not other historians of science, as their reference

---

[40] See the brilliant article by Mark Adams, who was the first to discuss this trend in Soviet science: Mark B. Adams, "Biology after Stalin: A Case Study," *Survey: A Journal of East and West Studies* 23 (1977–78): 53–80.

[41] For example, the Presidium of the Academy relocated to the Institute for the History of Science and Technology several scholars from the Institute of Philosophy and the Institute of General History who proved to be ideologically untrustworthy, including the former director of the Institute of Philosophy, the academician Bonifatii Kedrov.

group. This weakened the history of science but strengthened theoretical biology, which became a field of self-realization for the historians of biology.

Philosophers of biology were in a similar situation. Doing philosophy of natural sciences was more productive because it allowed for more innovation than standard Marxist dialectical materialsm and was somewhat grounded in actual science. It might have been less advantageous than engaging in sheer Marxist rhetoric, but it was also a lot safer. The Leningrad branch of the Institute for the History of Science and Technology at some point became a center for philosophers-turned-historians-of-biology, led by Kirill Zavadskii – botanist, evolutionary theorist, and active participant in the debates of the 1950s, who taught for a while in the Department of Philosophy of Leningrad University. In the 1970s, his research group focused on the history of evolutionary biology, organized several major conferences on evolutionary theory, and published subsequent collective volumes uniting biologists and historians of biology. These and many other publishing efforts (serial collective volumes in Tartu and Novosibirsk, many conference volumes, and the volumes of lectures from various summer schools) along with *Zhurnal Obshchei Biologii* (Journal of Biology) constituted a wide forum that made the field of theoretical biology rather vigorous.

In Russia, philosophers-turned-theoretical-biologists and scientists-turned-methodologists played an important role in shaping the field of theoretical biology. The field was constituted by the public positioning of actors against each other, and there were no shortage of scientists, philosophers, and historians willing to participate in this game. There were many groups and individual actors with different positions, interdefined in a struggle for peer recognition and authority. These groups often were in contact; their social circles did not necessarily intersect, but they nevertheless played on the same agonistic field. In short, theoretical biology in Russia was not a discipline with a disciplinary community but only an intellectual field, in Bourdieu's sense.

In the 1970s and 1980s, Shaposhnikov was an actor in this field, and his experimental results were widely discussed at the very center of the debates in Russian theoretical biology. We now turn to how these debates influenced his life and work.

SHAPOSHNIKOV BETWEEN ZOOLOGY AND THEORETICAL BIOLOGY

During the 1970s and 1980s, Shaposhnikov's life at the Zoological Institute was not easy. Beginning in 1961 he had been the head of a research group, supervising technicians and graduate students. In 1979, having reached the

age of retirement, he had to leave his position as a research fellow and become a consultant, but he was still paid some salary and was still considered
the head of his aphid research group.[42] In 1982, he was unexpectedly dismissed from his paid administrative position in the Zoological Institute on
the grounds of staff reduction, just when he felt himself to be at the peak
of his creativity and enjoyed recognition among evolutionary biologists for
his theoretical publications. After a few hard months of negotiations, the
Institute's authorities made a compromise, and Shaposhnikov was made an
adjunct researcher – a consultant without salary, which meant that he still had
his office and a graduate student to work with and still supervised the aphid
collections.

Many scientists who knew Shaposhnikov and his works were astonished
that he had been forced into retirement. Shaposhnikov's full retirement caused
various rumors and was immediately linked to the controversial reputation
of his research at the Zoological Institute. Nowadays, some scientists still
argue that Shaposhnikov had been considered in the Zoological Institute as
a Lamarckian, and that he was accordingly fired from the Institute for his
anti-Darwinian, even Lysenkoist views.[43] Since Shaposhnikov's experiments
were often used to disprove various Darwinist claims, some scientists are
convinced that Shaposhnikov's own Darwinian interpretations were merely a
public defense disguising his hidden Lamarckian views,[44] while others rather
furiously disclaim this judgment, insisting on their vision of Shaposhnikov as
a true Darwinian.

His experiments indeed fit well into different theoretical frameworks, but
the conflict with the administration had more to do with the general structure of scientific fields rather than with any particular position Shaposhnikov
occupied at any given moment. Shaposhnikov was both a taxonomist and a
theoretical biologist, and he spent a good part of his life trying to reconcile
the activities in these two fields. He succeeded well in converting his symbolic
capital as a taxonomist into capital in the field of theoretical biology, but he
failed to reconvert his authority as a theorist into a position in the field of
zoological taxonomy.

Microscopic plant lice are an obscure group and hard to work with, which
provided Shaposhnikov and his experiments with some symbolic capital in

---

[42] We have to stress that though there was an official retirement age, the institutes' administration
in the Academy had a great deal of freedom to keep researchers actually working long after
they officially retired.

[43] Interview with Alexander S. Rautian, Institute of Paleontology, Moscow, November 17, 1999.

[44] Yurii V. Chaikovskii, "Evoliutsiia: Part 5," 1 Sentiabria. Biologiia 43 (1997): 5–12.

the eyes of theoretical biologists. So great was his reputation as an expert on plant lice that nobody among theoretical biologists in Russia ever doubted the results of his experiments or his interpretation of them. His aphid experiments became popular among theorists precisely because they had that initial capital, and they accrued even more symbolic capital when various theorists began to refer to them in their arguments and realized that Shaposhnikov's plant lice could be used by all of the theoretical denominations to prove their own views.

The relation between the field of theoretical biology and the field of zoological taxonomy is asymmetrical; reputation as a theorist is hard to convert into reputation as a taxonomist. Most taxonomists work on obscure and difficult groups, and they are not fascinated when research on some minute insect becomes a cause of discord among various theoretical biologists. Theoretical biologists in Russia did not constitute a real community with shared norms, values, and communal practices – being totally disjointed and broken into groups, they simply formed an agonistic field of competition that anyone could enter. Entomological taxonomists, by contrast, for all their competition, formed a very tight community (with strong communal norms and values), which was hard to enter, and in which it was hard to gain a good reputation.

The symbolic capital of a taxonomist is in many respects determined by her or his collection. Shaposhnikov's position as head of the aphidological section in the reputed center of Russian zoological taxonomy, and his control over the largest collection of aphids – inherited from Mordvilko and enlarged greatly by Shaposhnikov himself – ensured him almost automatically the position of the leading aphidologist of both the Soviet Union and Eastern Europe. Unfortunately for him, he was unable to capitalize on this position.

Shaposhnikov's close colleagues in aphid studies respected his reputation as a theorist, but this activity was often considered a distraction from his basic mission. Once a colleague from Tajikistan responded bitterly to Shaposhnikov:

Frankly, I repeatedly reminded you to devote yourself, as a representative of the Zoological Institute, to the coordination of research activity in aphidology, integrating the specialists instead of allowing them to disperse. To give each of them the task of compiling the material and producing definite works on the different groups of aphids on the national scale. Alas, you didn't go in this direction. . . . If . . . we look at your activity at the center of aphidology in the Soviet Union we will see many mistakes and cases of the neglect of your duty: you were keen on your experiments with aphids transplantation and forgot your mission as the chief coordinator of aphidology.[45]

---

[45] M. N. Narzikulov to Shaposhnikov, April 29, 1983, Shaposhnikov papers.

In the Zoological Institute, scientific output was measured by the volume of taxonomic work – specifically, in the volumes published in the *Fauna of the USSR* series of fundamental monographs, each devoted to a certain systematic group. Shaposhnikov failed to produce one on aphids. Moreover, Shaposhnikov's activity in the field of theoretical biology marginalized him at the Zoological Institute: "Shaposhnikov was isolated at his institute. Many colleagues thought that his experimental and theoretical works had no connection to the work of the Institute, since instead of aphid description he devoted himself to theoretical problems."[46] His Czech student's recollections give a vivid picture of this isolation: "During my short stay in the former Leningrad in 1962, Georgiy invited me to listen to a lecture, which he gave in his laboratory adorned with diagrams and drawings of aphids. The subject of the lecture was 'Morphological divergence and convergence in aphids in the course of adaptation to an unusual host.' He read the lecture to me, as I was the only person present."[47]

Shaposhnikov's experiments and theories were also under suspicion at the institute, and his claims of having experimentally obtained a new species reminded many colleagues of Lysenko's times. Jaroslav Holman writes about Shaposhnikov in the 1960s that "Georgiy had obtained the results [on experimental speciation] about five years previously but had hesitated to make them known as he was afraid of their misinterpretation in Lysenkoan terms. . . . Initially I was skeptical of the validity of these results. My opinion changed gradually over the next few days when I checked the documentation and the slides."[48]

By contrast, Shaposhnikov's life with theoretical biologists was much more successful and rewarding. He visited many institutes around the Soviet Union and, as often happens with people marginalized in their home institutions, found them friendlier than his own Zoological Institute. The Paleontological Institute in Moscow was the most hospitable; there was always a friendly audience of people of different ages and research backgrounds willing to discuss his works from fresh and unorthodox perspectives. As one researcher recalled: "[The Paleontological Institute] was like a home for him. When he arrived all the people gathered specially to attend his lectures. It was always an event."[49]

---

[46] Interview with Lev Ia. Borkin of the Zoological Institute, St. Petersburg, October 7, 2000.
[47] Jaroslav Holman, "Reminiscences about Georgiy Shaposhnikov," in J. M. Nieto Nafria and A. F. G. Dixon, eds., *Aphids in Natural and Managed Ecosystems* (Leon: Universidad de Leon, 1998), pp. 675–6, at p. 675.
[48] Ibid.
[49] Interview with Rautian, October 26, 2000.

Shaposhnikov discussed his experiments among different audiences, attracting attention to them. These constant travels and presentations were in many respects for Shaposhnikov an attempt to gain legitimacy for his experimental research in the eyes of other zoologists and, at the same time, to draw attention to his subject among theoretical biologists. He never objected to any of the theoretical interpretations of his experiments, probably because opposing interpretations kept discussion of his work alive. By engaging in these activities, he was hoping to capitalize on his experiments, trying to raise symbolic capital by constant attempts to reconvert it between fields.

DIFFERENT INTERPRETATIONS OF SHAPOSHNIKOV'S EXPERIMENTS

We will now examine several debates in Russian theoretical and evolutionary biology that touched on Shaposhnikov's experiments, and in which different interpretations of Shaposhnikov's work were produced. This will allow us to review different intellectual positions in the field of theoretical biology in Russia during the 1970s and 1980s.

Before discussing the different interpretations of Shaposhnikkov's results, it is important to say a few words about the reception of these experiments by experts in aphidology. Shaposhnikov's experiments were known abroad, but initially, many of his foreign colleagues were very skeptical about the validity of his results, treating them as the result of possible contamination. This negative judgment tainted Shaposhnikov's reputation; he came to be known, not only as a bad experimentalist but also, first and foremost, as a taxonomist who couldn't distinguish his plant lice. The opportunity to clear his reputation and convince his foreign colleagues was given to Shaposhnikov at the First Aphidological Symposium, held in 1981 at Jablonna, Poland.[50] As one of

---

[50] Actually, Shaposhnikov himself was a main organizer of this meeting. As a good taxonomist, Shaposhnikov believed in the importance of international cooperation and personal communication and was dissatisfied that the few aphidologists were scattered across the world with opportunities for contact only at the huge entomological meetings. During the Thirteenth International Congress of Entomology (Moscow, 1968), Shaposhnikov organized an informal meeting of participants interested in aphids, which conceived the idea of an international aphidological symposium. Shaposhnikov was unsuccessful in organizing such symposia in Russia and therefore tried to encourage colleagues in other countries of the Soviet bloc. (See the recollections of the first steps in the organization of this symposium in Holman, "Reminiscences.") Nevertheless, Shaposhnikov remained one of the key figures at this symposium; his paper on the evolution of aphids, discussing his experiments in particular, opened the symposium, and his general comments closed the meeting. See Georgii Ch. Shaposhnikov, "The Main Features of the Evolution of Aphids," in *Evolution and Biosystematics of Aphids*

the organizers of this symposium, Shaposhnikov did his best to get the chief skeptics to attend. He brought to Jablonna all of his materials pertaining to the experiments – microscopic slides, notes, keys to identification of aphids – and ensured that there was a microscope available at the conference site, giving all doubters an opportunity to study the evidence. Seeing is believing: despite their initial doubts about the quality of the experiments, his colleagues, after having discussed this work with the author and seen Shaposhnikov's material, reached the general consensus that the "new form" described by Shaposhnikov in his earlier works was not the result of contamination.[51]

THE DISCUSSION OF RAPID SPECIATION

One of the hot spots in evolutionary discussions during the 1970s and 1980s was the possibility of rapid speciation. The synthetic theory of speciation considered evolution as a continuous adaptive process of the substitution of alleles at polymorphic loci, resulting in the gradual divergence of geographically separated populations, up to the species level. During the 1970s in the United States, an alternative viewpoint was proposed simultaneously by geneticists and paleontologists. On the one hand, some population geneticists were opposed to the standard theory of speciation within the evolutionary

(International Aphidological Symposium at Jablonna, 5–11 April, 1981) (Wroclaw: Ossolineum, 1984), pp. 19–99.

[51] Whatever one may think of Shaposhnikov's data and their interpretation, the only experts who could judge what plant lice Shaposhnikov got in his experiments considered the results clean and impressive. For the initial critical reception, see Roger L. Blackman, "Stability and Variation in Aphid Clonal Lineages," Biological Journal of the Linnean Society 11 (1979): 259–77; for a full acceptance, with apologies for initial doubts, see Roger L. Blackman, "Discussion after Shaposhnikov's Paper 'The Main Features of the Evolution of Aphids'," in Evolution and Biosystematics of Aphids, pp. 75–7, at p. 76; and Henry L. G. Stroyan, "Recent Development in the Taxonomic Study of the Genus Dysaphis," in Evolution and Biosystematics of Aphids, pp. 347–91, at p. 390.

This personal meeting in Jablonna and the ensuing change in opinion by leading experts on taxonomy and variations of aphids opened a new international vista, if not for Shaposhnikov, who never traveled widely, then at least for his experiments. Shaposhnikov's results, first regarded as sheer artifacts caused by violation of experimental standards, are now accepted as established and reliable scientific fact, which can be safely used by theoretical biologists. See, for example, John M Emlen, D. Carl Freeman, April Mills, and John H. Graham, "How Organisms Do the Right Things: The Attractor Hypothesis," Chaos 8 (1998): 717–26; Eva Jablonka, Beata Oborny, Istvan Molnar et al., "The Adaptive Advantage of Phenotypic Memory in Changing Environments," Philosohical Transactions of the Royal Society of London, series B – Biological Sciences, 350 (1995): 133–41; Csaba Pal and Istvan Miklos, "Epigenetic Inheritance, Genetic Assimilation and Speciation," Journal of Theoretical Biology 200 (1999): 19–37.

synthesis.[52] On the other hand, some paleontologists opposed to Simpson's and Mayr's views on evolution suggested the theory of "punctuated equilibrium."[53] Both geneticists and paleontologists implied the possibility of fast ("revolutionary") substantial evolutionary changes.

In Russia, the same trends appeared independently. Some geneticists in Russia came to conclusions analogous to Carson's and Powell's – in 1972, Yurii Altukhov and Yurii Rychkov published an article that claimed that speciation is not "a gradual process of changes expressed in terms of gene frequency dynamics but a qualitatively different phenomenon related to a rapid rearrangement of a part of the genome which is marked by the functionally most important monomorphic loci."[54] Later, Altukhov published two books in which he developed the same ideas.[55] These ideas were heavily criticized in Russia, where many scientists considered them a return to saltationist theories of evolution. In his later publications, Altukhov recognized that his views were somewhat similar to Eldredge's and Gould's, but did not find it worth the time and paper to discuss the theory of "punctuated equilibrium" proposed by paleontologists. He paid real attention only to the work of other geneticists – for example, Hampton Carson and Jeffrey Powell. He also never referred to Shaposhnikov and his experiments.[56]

---

[52] Hampton L. Carson, "Speciation as a Major Reorganization of Polygenic Balances," in C. Barigozzi, ed., *Mechanisms of Speciation* (New York: Alan R. Liss, 1982), pp. 411–33; Hampton L. Carson, "The Genetics of Speciation at the Diploid Level," *American Naturalist* 109, no. 1 (1975): 83–92; Jeffrey R. Powell, "The Founder-Flush Speciation Theory: An Experimental Approach," *Evolution* 32 (1978): 465–74.

[53] Niles Eldredge and Stephen J. Gould, "Punctuated Equilibrium: An Alternative to Phyletic Gradualism," in T. J. M. Schopf, ed., *Models in Paleontology* (San Francisco: Freeman Press, 1972), pp. 82–115.

[54] Yurii P. Altukhov and Yurii G. Rychkov, "Geneticheskii monomorfizm vida i ego biologicheskoe znachenie [The genetic monomorphism of species and its biological meaning]," *Zhurnal Obshchei Biologii* 33 (1972): 281–300; Yurii P. Altukhov, "Biochemical Population Genetics and Speciation," *Evolution* 36 (1982): 1168–81, at p. 1168. We quote the English article, but it is necessary to stress that our study is based mainly on Russian publications and analyzes mainly local reactions to them – publication in Russian journals played the most significant role in structuring the intellectual field.

[55] Yurii P. Altukhov, *Populiatsionnaia genetika ryb* [The populational genetics of fish] (Moscow: Pishchevaia promyshlennost', 1974); Yurii P. Altukhov, *Geneticheskie protsessy v populiatsiiakh* [Genetic processes in populations] (Moscow: Nauka, 1983). These works were heavily criticized in Russia; many scientists considered them a return to saltationist theories of evolution.

[56] It seems that the intellectual fields of genetics and theoretical biology were distinct, independent fields both in Russia and in the United States; in both countries, geneticists and paleontologists discussed similar problems quite independently without paying much attention to each other. For example, Hampton Carson in the United States, like Yurii Altukhov in Russia, never allied himself with the theory of punctuated equilibrium.

Russian paleontologists very early on were arguing against phyletic gradualism,[57] but the discussion never went in the direction of "punctuated equilibrium." When this idea reached Russia, it did not attract much attention among
paleontologists. By the time Eldredge and Gould introduced their concept,
Russian biologists had found their own "Russian" framework of analysis; they
discussed the problem of stasis and changes in the evolutionary process from
the vantage point of Schmalhausen's theory of stabilizing selection. Their interest was focused not on the explanation of stasis versus rapid change, but
on the explanation of rapid formation of the whole adaptive complex of characters. Both geneticists and paleontologists in Russia carried out studies in
this direction and independently arrived at the notion of destabilization, or
destabilizing selection.[58]

The geneticist Dmitrii Beliaev argued for the concept of destabilizing selection, which affected genetic homeostasis and produced an increase in variability. The variability caused by destabilizing selection then becomes the material
on which other kinds of selection act, thus accelerating the tempo of evolution.[59] Beliaev was interested in Shaposhnikov's works and communicated
with Shaposhnikov, but never cited him.

The paleontologist Zherikhin argued in 1967 that

the stabilizing state of variability is observed during continuous and monotonous evolution. In contrast, under conditions of environmental change, the previous adaptive
state became non-adaptive. . . . The state of destabilization is observed at the moment
of the conversion from one adaptive norm to another, accompanied by increased variability. Thus species are stabilized stages of the evolutionary process, divided in time
by phases of destabilization.[60]

Zherikhin's ideas gained popularity among his fellow paleontologists, and
all of them relied very much on Shaposhnikov's experiments as illustrations
of this idea:

In Shaposhnikov's experiments a sudden destabilization of many morphological characters was observed in conjunction with sharply increased level of elimination. The

[57] Vladimir V. Zherikhin, "Deiatel'nost' mezhsektsionnogo seminara po problemam evolitsii
[The work of the intersectional seminar on evolution]," *Biulleten' MOIP Otdelenie Biologiia*
72, no. 4 (1967): 136–8.
[58] Zherikhin, "Deiatel'nost' mezhsektsionnogo seminara"; Dmitrii K. Beliaev, "O nekotorykh
voprosakh stabiliziruiushchego i destabiliziruiushchego otbora [On some aspects of stabilizing
and destabilizing selection]," in K. M. Zavadskii, ed., *Istoriia i teoriia evolutsionnogo ucheniia*
(Leningrad, 1974), pp. 76–84; Dmitrii K. Beliaev, "Destabilizing Selection as a Factor in
Domestication," *Journal of Heredity* 70 (1979): 301–8.
[59] Beliaev, "O nekotorykh voprosakh."
[60] Zherikhin, "Deiatel'nost' mezhsektsionnogo seminara."

phase of destabilization was followed by the phase of stabilization of the new adaptive norm, accompanied by adaptation to the new host plant and by the decrease of level of elimination to control level. The new population had the morphological characteristics of a good species and was practically totally reproductively isolated from the ancestral population.[61]

During the 1970s and 1980s, this concept of "evolutionary destabilization," which was considered a further development of Schmalhausen's theories, became popular among Russian evolutionary biologists. However, these developments had taken different routes in different disciplinary groups of scientists. Paleontologists and (often) geneticists were the most liberal in their deviation from the orthodoxy of synthetic theory. Many scientists kept in mind the Lysenko period and preferred to be very cautious in their theorizing, guarding Russian biology from any traces of what they considered "Lamarckist reasoning." Some, in discussing problems of speciation from the viewpoint of Schmalhausen's theory and arguing that it occurs with a switch in development from one adaptive norm to another, were careful to state that they did "not have in view the resurrection of Lamarckism."[62] Such remarks were not the unnecessary remnants of old fears. At that time, Lamarckian explanations of adaptive modifications were still being expressed by Russian scientists, and Shaposhnikov's experiments were again at the center of the recurring controversies.

THE DISCUSSION OF LAMARCKISM

For Russian Lamarckians during Lysenko's reign and long afterward, Shaposhnikov's plant lice were favorite examples to be cited as proof of direct inherited environmental influence on the organism. For example, Evgenii Smirnov, defending Lamarckism and the inheritance of acquired characteristics both during and after the time of Lysenkoism, did similar experiments with aphids a few years before Shaposhnikov. He believed that his own and Shaposhnikov's results were good demonstrations of Lamarckian adaptations. As he wrote to Shaposhnikov at the beginning of Shaposhnikov's work: "I learned with great pleasure that your work goes well. It pleased me especially

---

[61] Aleksandr S. Rautian, "Paleontologiia kak istochnik svedenii o zakonomernostiakh i faktorakh evolutsii [Paleontology as a source of knowledge on the factors and laws of evolution]," in V. V. Menner and V. P. Makridin, eds., *Sovremennaia paleontologiia*, vol. 2 (Moskva: Nedra, 1988), pp. 78–117, at p. 89.

[62] Boris M. Mednikov, "Problema vidoobrazovaniia i adaptivnye normy [The problem of speciation and the adaptive norms]," *Zhurnal Obshchei Biologii* 48 (1987): 15–26, at p. 21.

in view of the fact, that the reviving Morganists want to devalue all our efforts. In this situation perfectly performed experiments are the best response."[63]

Much later, in the 1970s, another persistent Lamarckian and one of the leading Russian paleontologists, Leo Davitashvili, became very interested in Shaposhnikov's experiments. In his article, published in *Voprosy Filosofii* (Journal of Philosophy), he opposed "post-neo-Darwinism," as he called the synthetic theory, and cited Shaposhnikov's experiments to support his position.[64] A year earlier, he had made the same claim in the English publication *Evolution*, arguing that "the synthetic theory of evolution is not able to solve the main difficulties it is faced to."[65] Citing Shaposhnikov's experiments as an example of "hereditary adaptation to the new kind of food after just a few generations in a short period of time (one season)," he concluded, "I could not understand this case until I admit a hereditary change resulting from the direct action of the new kind of food."[66]

In his huge two-volume treatise *The Doctrine of Evolution*, published a few years later, Davitashvili devoted a substantial part of his discussion to experiments on food adaptation in insects, concluding that the Baldwin effect couldn't explain the fast origin of new hereditary characters and that such cases should be considered as an unquestionable inheritance of acquired characters.[67] Among others, he considered the experiments of Smirnov and Shaposhnikov: "For the unprejudiced reader, the results of the experiments carried out by Smirnov and his colleagues are beyond doubt, but it is good that these experiments were continued by a scientist who cannot be suspected of Lamarckist sympathies, sharing in his last works the positions of the 'synthetic theory of evolution' – G. Shaposhnikov."[68]

Davitashvili, having met Shaposhnikov personally, discussed the experiments with him and was much disappointed when Shaposhnikov disagreed with his interpretation. As Davitashvili wrote to Evgenii Smirnov:

I am very interested on the facts of inheritance of 'acquired characters', especially acquired adaptations, in particular – the hereditary changes of plant lice, described by

---

[63] Evgenii Smirnov to Shaposhnikov, February 10, 1957, Shaposhnikov papers.

[64] Leo Sh. Davitashvili, "Postneodarwinism i darwinism [Post-neo-Darwinism and Darwinism]," *Voprosy Filosofii* 1 (1970): 122–30.

[65] Leo Sh. Davitashvili, "Deficiencies of the Synthetic Theory of Evolution," *Evolution* 23 (1969): 513–16, at p. 513. Here we use the English paper for the citation, but in the Russian intellectual field the appearance of Davitashvili's article in *Voprosy Filosofii* [Journal of Philosophy] was much more important.

[66] Ibid., p. 514.

[67] Leo Sh. Davitashvili, *Evolutsionnoe uchenie* [The evolutionary concept], 2 vols. (Tbilisi: Metsniereba, 1977).

[68] Ibid., vol. 1, p. 119.

Shaposhnikov. This scientist tried to explain the facts in the spirit of classical genetics. When I got acquainted with him, he told me that you interpret these facts as hereditary changes, induced by the influence of environment and food. I replied to him frankly that in my opinion there can't be any other explanation. Georgii Shaposhnikov, in my view, is a very intelligent, serious scientist, but I suspect him to be under the strong influence of his institutional environment.[69]

Evgenii Smirnov replied to this letter: "Concerning Shaposhnikov, you are quite right: obtaining remarkable results, he doesn't name things by their names, because he doesn't want to break the canons."[70] It should be empha-sized that though Shaposhnikov's position seemed unorthodox to opponents of the synthetic theory, for those who knew Shaposhnikov it was clear that he never was a heretic, precisely because he didn't want to break the canons.

Shaposhnikov's experiments got additional attention in a popular jour-nal, where two authors opened a discussion devoted to the experiments. One of them, Alexei Jablokov, well known as an orthodox supporter of the syn-thetic theory of evolution, chose to downplay them; he was disturbed by Davitashvili's writings, which seemed to him dangerous, and stood guard against attempts to interpret Shaposhnikov's experiments in a Lamarckian sense.[71] The other, the anti-Darwinian paleontologist Sergei Meyen, argued, by contrast, that Shaposhnikov's experiments showed the possibility of "so-matic induction," or at least left the question open.[72] It is significant that neither author was expert in the issues he discussed. Meyen never touched questions related to Lamarckism in his academic publications – that is, in his articles in scientific periodicals. His publication in a popular scientific magazine reflects his intellectual position in the field of evolutionary biol-ogy rather than any Lamarckian sympathies. This discussion in a popular magazine also shows the prevalence of the logic of the field: an orthodox position taken by one protagonist called for an opposing heterodox position by another, and the plant lice were used as a device for interdefining these positions and strengthening synthetic orthodoxy. Both Jablokov and Meyen appealed to Shaposhnikov in private. Jablokov warned him that his reputa-tion could be damaged by Davitashvili's use of Shaposhnikov's work, arguing that "your publication with the clarification of all these matters is quite nec-essary."[73] Shaposhnikov, characteristically, never published any clarification. Meyen's letter evidently was never even answered. It seems that as much as

---

[69] Leo Sh. Davitashvili to Evgenii Smirnov, February 3, 1968, ARAN, f. 2079.

[70] Smirnov to Davitashvili, February 25, 1968, ARAN, f. 2079.

[71] Alexey V. Iablokov, "Zakon! Est' Zakon!," *Znanie–Sila* 9 (1974): 9.

[72] Sergei V. Meyen, "Zakon? Est' zakon?," *Znanie–Sila* 9 (1974): 8–9, p. 8.

[73] Alexei V. Iablokov to Shaposhnikov, March 26, 1970, Shaposhnikov papers.

other scientists wanted Shaposhnikov to appear heretical, he himself shied away from taking any radical stand.

THE DISCUSSION OF EPIGENETIC EVOLUTION

As we have seen, the intellectual field of theoretical and evolutionary biology during the 1970s and 1980s had virtually the same composition that it had had forty years earlier. Our case study reveals the same three main intellectual positions regarding the experimental studies of adaptive modifications as were held in the 1930s and 1940s: Lamarckian, orthodox Darwinian, and epigenetic approaches. This third intellectual position could be called "epigenetic Darwinism," and it was also projected onto Shaposhnikov's experiments.

The paleontologist Mikhail Shishkin's interpretation of Shaposhnikov's results is a good example of this third line of theorizing in evolutionary biology. In the 1980s, Shishkin developed Schmalhausen's theory of stabilizing selection and C. H. Waddington's concept of "epigenetic landscape" into what he called the "epigenetic theory of evolution." One of his statements was that the division between hereditary and nonhereditary variability (in other words, between mutations and modifications) is merely an abstraction. Instead of such a division, he proposed distinguishing the different realizations of the same genotype in different developmental conditions. The main problem of Darwinian theory, Shishkin argued, is the question of the causes of stable ontogenetic development resulting in the adaptive norm. Not heredity and its variation, but ontogeny and its variation, is the source of evolutionary change – hence the "epigenetic theory" of evolution.[74]

The problem of the inheritance of acquired characters is reformulated in this approach and replaced by another problem. Instead of asking about the hereditary acquisition of changes, Shishkin suggested asking, "Can an organism's particular unstable reaction become stable and constant in its progeny? Lamarckians are quite right in responding affirmatively to this question; their mistake begins in their explanation of how it occurs. The Darwinian theory of stabilizing selection also gives an affirmative response [to this question]," but provides a different causal explanation.[75]

---

[74] Mikhail A. Shishkin, "Individual'noe razvitie i estestvennyi otbor [The ontogenesis and the natural selection]," *Ontogenez* 15 (1984): 115–36; Mikhail A. Shishkin, "Evolutsiia kak epigeneticheskii protsess [Evolution as an epigenetic process]," in V. V. Menner and V. P. Makridin, eds., *Sovremennaia paleontologiia*, vol. 2 (Moskva: Nedra, 1988), pp. 142–68.

[75] Mikhail A. Shishkin, "Fenotipicheskie reaktsii i evolutsionnij protsess (eshche raz ob evolutsionnoj roli modificactij) [Phenotipic reactions and the evolutionary process (once more on the evolutionary role of adaptive modifications)]," in Ya. M. Gall, ed., *Ekologiia i evolutsionnaia teoriia* (Leningrad: Nauka, 1984), pp. 196–216, at p. 207.

Indeed, Shishkin also subscribed to the idea of stable and destabilized periods in species history. He often referred to Shaposhnikov's results, regarding them as the best proof of the significance of epigenetic transformations in evolution, denied by the synthetic theory: "The experiments by Shaposhnikov give the most obvious experimental confirmation of such a course of transformation: . . . changes in the developmental milieu lead to the realization of extremely unstable physiological states, the most viable of which turn into stable adaptation by means of selection."[76]

Shishkin widely publicized his concept and greatly contributed to the reestablishment of Schmalhausen's epistemic position in theoretical biology. After a temporary absence during the 1950s and 1960s, when scientists were attracted by the struggle with Lysenko and by the promotion of the synthetic theory of evolution in Russia, this position came back with the full revival of the intellectual field of Russian evolutionary biology during the 1970s and 1980s. Combining epigenetic theory with Darwinism, this line of theorizing became an advantageous intellectual position for many scientists who believed the synthetic theory to be too narrow but who at the same time abhorred Lamarckism as such, especially after the blackening of its reputation by Lysenko. History had come full circle: the position that had been dominant in the 1940s once again gained, if not dominance, then at least substantial weight forty years later.

Shaposhnikov, who almost never agreed directly with proposed interpretations of his work, entirely accepted Shishkin's interpretation of his experiments. Finally, after all the theoretical picking and choosing, at the epicenter of clashes between different positions over his experiments, Shaposhnikov subscribed to the Schmalhausen-like theory of epigenetic evolution. Soon after they met, Shaposhnikov wrote to Shishkin: ". . . I came to the conclusion that I accept your concept entirely. Apparently I was prepared by my respective reflections on the results of my experiments and by some other factors. Nevertheless, complete acceptance was hard, the inertia of thought is too strong. I can imagine how your concept will be assimilated by those who have repeated too many times the classical dogmas of population genetics."[77]

CONCLUSION

As we have attempted to show, the theoretical debates in Russian evolutionary biology have displayed substantial continuity from the late 1930s to the late 1980s. Debates about various theoretical issues continued even under

---

[76] Shishkin, "Evolutsiia," p. 160.
[77] Shaposhnikov to Mihail A. Shishkin, October 18, 1981, Shaposhnikov papers.

Lysenko, interrupted for only a few years: from 1948 to 1952, when they began to resurface, focusing on species concepts. Moreover, opposition to Lysenkoism gave vigor and appeal to theoretical debates in biology in general, and to new evolutionary concepts and the synthetic evolutionary theory in particular.

It seems that Bourdieu's concept of the "intellectual field" well fits such an endeavor as Russian theoretical biology, which was neither an established discipline populated by scientists with specific disciplinary identities, nor an interdisciplinary research area with no "native inhabitants." It had only relative independence from other fields, but Shaposhnikov's case demonstrates that it had specific capital of its own. Shaposhnikov had two distinct reputations (two types of symbolic capital, in Bourdieu's terms): in entomology and in theoretical biology. Shaposhnikov's failure to reconvert his reputation in theoretical biology into a position in taxonomy proves better than anything else the specificity of "theoretical biology capital" and the independence of this field.[78]

Shaposhnikov's fate shows us the irony of a good scientist's trying to be original and orthodox at the same time. The epistemic position that Shaposhnikov subscribed to in adopting the synthetic theory during his early years was rather unorthodox at the time, and adopting the Schmalhausen-like approach during a period when the synthetic theory was generally recognized and accepted in Russia as the ultimate answer to all evolutionary questions was likewise rather unorthodox. In order to appreciate fully the contextual meaning of orthodoxy, we may compare the epistemic position of epigenetic evolution in three national variants of the scientific field of evolutionary biology: Russian, American, and British.

Such a comparison is possible owing to the fact that two notorious figures, I. Schmalhausen and C. H. Waddington, presented homologous epistemic positions in homologous intellectual fields. In the 1940s in England, simultaneous with Schmalhausen, Waddington proposed a concept very close to Schmalhausen's – "canalizing selection" and the "genetic assimilation" of adaptive characters.[79] Like Schmalhausen, Waddington united the epigenetic approach with the theory of natural selection. Waddington evidently

---

[78] According to Bourdieu, the different kinds of capital can be reconverted to each other (political capital into scientific authority, or the specific capital of one scientific field into the specific capital of another), but only the inability of an actor to reconvert capital proves the independence of the field and the specificity of the capital itself.

[79] See, for example, Conrad H. Waddington, "Canalization of Development and the Inheritance of Acquired Characters," *Nature* 3811 (1942): 563–5.

considered Schmalhausen his main competitor;[80] Schmalhausen likewise considered Waddington his competitor. In 1944, after becoming acquainted with Waddington's works, Schmalhausen made an attempt to publish his main book, *The Organism as a Whole in Development and Evolution*, abroad. He expected that Julian Huxley would assist him. As he wrote to Huxley: "I should be happy indeed were you personally to take upon yourself the editing of the translation.... [My book] is quite unknown abroad.... Meanwhile, various other authors are putting forward hypotheses analogous to those elaborated by myself. It is only now, on my return to Moscow after two years evacuation, that I have read C. H. Waddington's article in 'Nature' (end of 1942). Our opinions are in essence identical, the terminology alone differing (I use the expression 'autoregulation' instead of 'canalisation')....I am endeavoring to send...my other book 'Factors of evolution' to America where apparently there are greater facilities for the publication of such books."[81]

Waddington's and Schmalhausen's similar concepts met similar chilly receptions from leading American evolutionary biologists. From the point of view of American classics of the evolutionary synthesis, both Schmalhausen and Waddington appeared to be rather deviant. In spite of Dobzhansky's endeavours to promote Schmalhausen's work, its reception was by no means enthusiastic. In his generally benevolent review of Schmalhausen's book, G. G. Simpson nevertheless allowed the relevance of this approach only to Russia, as a valid compromise position for Russian geneticists and Darwinians

---

[80] See the exchange between Waddington and Dobzhansky on Schmalhausen in Conrad H. Waddington, *The Evolution of an Evolutionist* (Edinburgh: Edinburgh University Press, 1975).

[81] Schmalhausen wrote this letter in English, and we quote it in its original orthography: Ivan I. Schmalhausen to Julian S. Huxley, Schmalhausen papers (ARAN, f. 1504, inv. 3, folder 9). However, an English edition of Schmalhausen's first book was never published. Only in 1949, with Dobzhansky's help, was Schmalhausen's *Factors of Evolution* published in America: Ivan I. Schmalhausen, *Factors of Evolution: The Theory of Stabilizing Selection*, trans. I. Dordick, edited and with a Foreword by Th. Dobzhansky (Philadelphia and Toronto: Blakiston, 1949). Schmalhausen always emphasized his independence and his priority in the development of this direction of thought. He wrote to Waddington in 1962: "The idea of the significance of the stabilizing form of selection occurred to me long ago, and I have been cherishing it for almost 20 years....I am especially glad that you are trying to find direct proofs and to check *our* views experimentally on genetically well studied material" (our emphasis). (Ivan I. Schmalhausen to Conrad H. Waddington, [1962], Schmalhausen papers [ARAN, f. 1504, inv. 3, folder 4]). Obviously, Waddington didn't reply to this letter, and, even more strikingly, he avoided meeting Schmalhausen during his October 1962 stay in Moscow, not going beyond a short formal letter mailed to Schmalhausen (Conrad H. Waddington to Ivan I. Schmalhausen, October 19, 1962, Schmalhausen papers (ARAN, f. 1504, inv. 3, folder 73).

suppressed by the dictatorship of Lysenko.[82] The reception of Waddington was even worse. Both Mayr and Dobzhansky (in his later works) criticized Waddington's concept, claiming that Waddington's term "genetic assimilation" was poorly chosen and interpreting Waddington's theory as a failed attempt to support Lamarckian inheritance.[83]

We would argue that the independent emergence of similar views was determined by the similar dynamics of these two fields. The field of theoretical and evolutionary biology in Russia seems to have been similar to the American and British fields, though in America theoretical biology was less well established than in Britain or Russia. It is worth noting, for example, that in both Russia and America paleontologists happened to be at the center of theoretical activity during the 1970s, which cannot be explained simply by reducing the problem to personalities such as Stephen Jay Gould and Vladimir Zherikhin. It seems there must have been some social cause for this activity, probably in the relation of the field of paleontology to other fields and in the specific symbolic capital that paleontologists had to spend in theoretical biology. At the same time, the resemblance of Waddington's and Schmalhausen's concepts and their dominance in both Russia and England during the 1940s, as well as the similarity of their receptions by American evolutionary biologists,

[82] As Simpson wrote in his review: "Followers of his [Schmalhausen's] theory might... accept the evidence of the neo-Lamarckians or of Lysenko at face value, and still accept also the full findings of orthodox genetics. The popularity and intense cultivation of this solution occurred in the Soviet Union in the period of struggle between the Mendel-Morganists and the Michurinists" (G. G. Simpson, "'Factors of Evolution,' a Review," *The Journal of Heredity* 40 [1949]: 322–4, at p. 323). Simpson denied the relevance of the very notion of "stabilizing selection":

> Stabilizing selection in the narrow sense ... is doubtless a real phenomenon. It seems to have been established experimentally by Gause and to be the most probable explanation of some observed cases of adaptation. It is also of considerable interest in its bearing on some of the test cases of Neo-Lamarckism. There is, however, room for question whether it merits such extreme emphasis, aside from the ideological struggle which evidently stimulated some of these studies. (ibid.)

[83] Canalizing selection tends to stabilize the development pattern of a species, to make certain traits develop similarly in most environments that this species encounters in its habitats. The result is eventually that some traits show little or no variation, being uniform in all representatives of the species.... Waddington's (1953) so-called 'genetic assimilation of an acquired character' is a similar tour de force, but achieved by manipulation of the external rather than of the genetic environment.... The analogy with alleged Lamarckian inheritance is superficial.... [Canalisation] is evidently a result of developmental buffering, and not of lack of genotypic variance. (Dobzhansky, *Genetics of the Evolutionary Process*, p. 211) See also Scott F. Gilbert, "Induction and the Origins of Developmental Genetics," in Scott F. Gilbert, ed., *A Conceptual History of Modern Embryology* (Baltimore and London: Johns Hopkins University Press, 1991), pp. 181–206, at p. 205.

demonstrates that in Britain and Russia, in contrast to the U.S., the existence of extreme views as legitimate positions within the field (for example, openly professed Lamarckism and rigorous neo-Darwinism, or holism and reductionism) widened the space for scientists to put forth their claims and define their positions somewhere between the extremes – that is, between heresy and orthodoxy. Schmalhausen's theory in Russia and Waddington's theory in Britain were strategies for gaining dominance by taking the middle ground between polar theoretical positions. They appealed to many scientists on all the sides of the debates. It seems that America was different from Europe in one important respect – in America, vitalists and Lamarckians were much more effectively elbowed out of the field of evolutionary biology very early on, and thus in America there was neither a need nor a place for the Schmalhausen–Waddington intellectual strategy.

In the context of Russian theoretical biology, Georgii Shaposhnikov (like Ivan Schmalhausen, in fact) was both orthodox and a heretic, following William of Baskerville's dictum from *The Name of the Rose* cited in the epigraph. He tried to gain theoretical prominence and originality by sticking to a canon less popular than the central orthodoxy of his time. Many theoreticians from all camps reviewed his work and appreciated his views, though most of them wanted him to be more radical, indeed cajoling him to join their sides. He agreed with almost everyone, and joined no one, trying to find a "golden middle" which would suit him and his intellectual ambitions. It is exactly this strategy that makes Shaposhnikov such an interesting case in the history of Russian biology and allows us from its vantage point to view the development of the whole field of evolutionary and theoretical biology in Russia over the last half century.

CHAPTER THREE

The Specter of Darwinism

The Popular Image of Darwinism in Early Twentieth-Century Britain

Peter J. Bowler

It's been a long time now since I borrowed the phrase "the eclipse of Darwinism" from Julian Huxley's 1942 survey *Evolution: The Modern Synthesis* for the title of my own book on the non-Darwinian evolutionary theories that proliferated around 1900.[1] But even at the time, I was conscious of a problem with Huxley's use of the term "eclipse" to denote the state of Darwinism during the period before the emergence of the genetical theory of natural selection. An eclipse is a temporary diminution of brightness, implying that Darwinism had gained considerable influence in science during the period following the publication of the *Origin of Species*, before being overtaken by a temporary wave of opposition at the turn of the century. Yet all the work that historians have done on the origins of evolutionism seems to suggest that the theory of natural selection encountered massive opposition from the start, and that Darwin had succeeded in popularizing the basic idea of evolution even though few scientists (or anyone else, for that matter) thought that his explanation of its workings was adequate. Even Julian's grandfather Thomas Henry Huxley, popularly known as "Darwin's bulldog," turns out on closer inspection to have been lukewarm in his support for the selection theory and inclined to look for alternatives, such as saltationism.[2]

---

[1] Peter J. Bowler, *The Eclipse of Darwinism: Anti-Darwinian Evolution Theories in the Decades around 1900* (Baltimore: Johns Hopkins University Press, 1983); see also Julian Huxley, *Evolution: The Modern Synthesis* (London: Allen and Unwin, 1942), p. 22.

[2] Michael Bartholomew, "Huxley's Defence of Darwinism," *Annals of Science* 32 (1973): 525–35; Mario Di Gregorio, *T. H. Huxley's Place in Natural Science* (New Haven, CT: Yale University

Why, then, did Julian Huxley use a phrase that implied that Darwinism *had* been popular? One simple answer is that he was exploiting an ambiguity permitted by the changing meaning of the term "Darwinism." No one denies that something called "Darwinism" was popular during the 1870s and 1880s, and it would have been easy to imply that this early form of Darwinism was equivalent to the "neo-Darwinism" of the late nineteenth century, which evolved into the Darwinism of the Modern Synthesis – that is, a reliance on natural selection as the dominant mechanism of evolution, with little or no input from the more teleological processes invoked by some anti-Darwinians. In fact, this early "Darwinism" was really only evolutionism broadly conceived – only the most extreme exponents of anti-Darwinian models were excluded, and most Darwinians would have been even more liberal than Darwin himself in allowing a role for mechanisms other than natural selection. Julian Huxley realized that much of the opposition generated among younger biologists during the 1890s was directed against the phylogenetic reseach associated with the original Darwinian program, something that was only loosely associated with the selection theory. But he knew that the opposition had spilled over into hostility toward the selection theory itself, and by using the term "eclipse of Darwinism," he encouraged a belief that selection had suffered a temporary loss of influence. Recent research on the relationship between science and religion in early twentieth-century Britain has suggested that there is more to it than this, however. It is now clear to me that Huxley was articulating a general feeling among his contemporaries that the selection theory had at first gained massive support within nineteenth century science by riding the coattails of a more broadly based materialistic philosophy. A wide range of theologians and popular writers interested in the implications of science promoted the assumption that materialism in general, and Darwinism in particular, had become dominant forces in Western thought before being self-consciously repudiated during the period around 1900. Most of these people believed that Darwinism was now completely dead and could never be revived – in this sense, Huxley was promoting a significant new development that many in his generation had not anticipated: the renewed interest in the selection theory made possible by the claim that Mendelian genetics left that mechanism as the only plausible theory of evolution.

My study has generated a number of questions relating not so much to the development of science itself but to the perception of science by nonscientists,

Press, 1984); more generally, see Peter J. Bowler, *The Non-Darwinian Revolution: Reinterpreting a Historical Myth* (Baltimore: Johns Hopkins University Press, 1988).

including theologians and popular writers on philosophical topics. A whole
generation of British churchmen were convinced that science, including biology, had turned its back on the materialism of the nineteenth century and
was now once again ripe for a synthesis with liberal Christianity via the creation of a new natural theology based on progressionist evolutionism, the
reintroduction of teleology, and a nonmaterialistic physiology. Their views
were shared by popular writers such as Bernard Shaw, who were striving to
promote a new philosophy of life that was non-Christian but that had many
similarities to the more liberal Christian viewpoint of the "Modernists" within
the churches. These people lapped up information about the revival of vitalism in physiology and the increasingly articulate preference of evolutionists
for non-Darwinian theories. They were convinced that science was moving
toward a non-materialistic and teleological worldview with which liberal religious thinkers could do business. Although aware that some anti-Darwinians
promoted an experimentalist viewpoint against the old natural history tradition, their real interest was in the older form of anti-Darwinism based on
Lamarckism and other processes that were thought to allow a role for purpose
in evolution.

Huxley was closely in touch with this movement, for although he rejected
the idea of a personal God, he strove to retain a place for the religious feelings
in his philosophy and remained friendly with a number of liberal churchmen.[3]
Yet outside the small circle of those who were in direct contact with working
scientists, most clergymen and nonspecialist writers got their information
about science from the popular writings of a few leading scientists – a genre
of literature dominated by older figures who had been active at the end of the
nineteenth century but who became increasingly out of touch with research
developments as the twentieth century progressed. A substantial proportion
of the general public was thus given a false impression of the true direction of
contemporary science by a small and unrepresentative group of writers who
had good access to the media. The episode offers interesting lessons about the
interaction between scientists and the public and may even offer a warning that
we need to be on our guard against the possibility of similar misperceptions
in the present.

---

[3] Julian Huxley, *Religion without Revelation* (London: Ernest Benn, 1927). On Huxley's life and
opinions, see, for instance, Kenneth C. Waters and Albert Van Helden, eds., *Julian Huxley:
Biologist and Statesman of Science* (Houston: Rice University Press, 1992). Huxley was friendly
with William Temple (the future archbishop of Canterbury) and sometimes met for discussions
with a group led by the Modernist theologian B. H. Streeter; on his involvement in religious
debates, see Peter J. Bowler, *Reconciling Science and Religion: The Debate in Early Twentieth-
Century Britain* (Chicago: University of Chicago Press, 2001), Chapter 4.

Equally interesting to the historian is the fact that in the course of creating this unrealistic image of contemporary science, the antimaterialist lobby also succeeded in setting up a distorted image of late nineteenth-century science as totally dominated by materialism. This distortion may have taken in Julian Huxley himself – hence the "eclipse of Darwinism" metaphor – and it continued to mislead historians of science until dispelled by new work that revealed the true nature of the initial form of Darwinism by showing how much of it represented a compromise with earlier ways of thought. Examples will be given later to show that there were many who thought that the eclipse – or the death – of Darwinism represented the collapse of a substantial commitment to materialism and the selection theory by the scientific community of the previous generation. The dire warnings of the antimaterialists about the danger from which science had so narrowly escaped give the impression that they were only too anxious to exaggerate the earlier level of support for the selection theory. Here we see another example of the baleful influence of the metaphor of the "war" between science and religion – a metaphor that seems to appeal to extremists on both sides, unwilling as they always are to admit that there really is a substantial middle ground in the field of opinion. To some extent, this metaphor was created by supporters of T. H. Huxley's philosophy of scientific naturalism, which found its classic expression in J. W. Draper's *History of the Conflict between Religion and Science* of 1874.[4] But just as it suited the most enthusiastic advocates of naturalism to present it as a force that would sweep religion away, it suited their opponents of the next generation to pretend that they were breaking up a dogma that had been accepted quite uncritically because it fitted a materialist ideology that had threatened the very foundations of Western culture. By magnifying the dimensions of the crisis, they highlighted the significance of the "new" synthesis of science and liberal religion that they wished to proclaim. In fact, of course, the anti-Darwinian theories on which they relied were hangovers from the very period that they sought to depict as riddled with materialism. Just as the upsurge of interest in nonmaterialist physiology at the end of the nineteenth century proved to be only temporary, these theories were already becoming outdated during the first decades of the twentieth. The new anti-Darwinian theories based on experimentalism and Mendelian genetics were just as hostile to the old worldview as was the Darwinian specter invoked by the critics. This is why the old materialism and the new were already coming together to create the

---

[4] For a critique of the warfare metaphor, see James R. Moore, *The Post-Darwinian Controversies: A Study of the Protestant Struggle to Come to Terms with Darwin in Great Britain and America, 1870–1900* (Cambridge: Cambridge University Press, 1979).

foundations of the Modern Synthesis while the critics were still desperately trying to convince themselves that the supposed decline of Darwinism was a prelude to its extinction.

In order to put flesh on the bones of this argument, I have divided this chapter into three parts. The first looks at the changing meaning of the term "Darwinism" around 1900 and suggests that it was increasingly being used in the narrow sense denoting only the theory of natural selection. The second section provides examples of early twentieth-century writers who not only used the term in this narrower sense, but also insisted that they were witnessing the collapse of an all-powerful Darwinism that had been accepted as dogma by an earlier generation of scientists blinded by materialism. The final section very briefly describes the sources of the non-Darwinian models that were being proclaimed as evidence that science had indeed turned its back on materialism. With hindsight, of course, we can see that this was a completely unrealistic picture of the state of contemporary biology. When Huxley and J. B. S. Haldane proclaimed the reprieve of Darwinism from its supposed death sentence, they were providing the first (and initially not very successful) efforts by the younger generation of scientists to counter this image in the popular imagination.

THE CHANGING MEANING OF "DARWINISM"

By the 1870s, what we now call the theory of evolution was securely established, although few believed that Darwin's natural selection provided a complete explanation of how the process worked. The general idea was known as the "doctrine of descent" and incresingly as the "theory of evolution."[5] But the term "Darwinism" was also in use, acknowledging Darwin's position as the figurehead of the campaign to establish the theory of evolution. There is little evidence of that term's being used in the narrow sense, in part because Darwin himself insisted that he did not rule out a role for other mechanisms, such as the inheritance of acquired characteristics. By the last decade of the century, however, identification of Darwin with the theory of natural selection, and use of the term "Darwinism" to denote a narrow selectionism, became increasingly common. The wider meaning did not disappear completely – as late as 1935, Arthur Keith insisted that in defending "Darwinism" against its critics he was only defending the general theory of evolution (and Keith himself was no rigid selectionist).[6] But this was an unusual tactic – by this

---

[5] See my "The Changing Meaning of 'Evolution,'" *Journal of the History of Ideas* 36 (1975): 95–114.

[6] Arthur Keith, *Darwinism and Its Critics* (London: Watts, 1935), p. 3.

time, most writers would have used "Darwinism" in its narrower sense and would have linked its rise and fall with the changing fortunes of materialism.

In the second volume of his *Darwin and After Darwin*, published after his death in 1894, George John Romanes contrasted the original, flexible form of Darwinism with the rigid selectionism now being advocated by Alfred Russel Wallace and August Weismann. He suggested the use of the terms "neo-Darwinism" and "neo-Lamarckism" to denote the two schools created by the fragmentation of the synthesis contained in Darwin's own work.[7] He even tried to float the term "Wallaceism," but this flew in the face of the fact that Wallace's own *Darwinism* of 1889 had, with characteristic generosity, used Darwin's name to denote the theory of evolution by natural selection.[8] With the exception of human origins, Wallace was a rigid selectionist, and his use of the term "Darwinism" to denote this position preempted Romanes's "neo-Darwinism" and marks the beginning of a trend toward using the term in the narrow sense, excluding Lamarckism or any other teleological process. Samuel Butler also mentioned the term "Wallaceism," but the title of his 1890 article "The Deadlock in Darwinism" makes it clear that he expected his readers to associate "Darwinism" with the selection theory.[9]

By the early twentieth century, the identification of Darwin and Darwinism with the selection theory was becoming widespread in both Britain and America. William Bateson attacked Darwin's theory of natural selection (and emphasized that it was Darwin's) in his *Materials for the Study of Variation* of 1894, although he did not use the term Darwinism.[10] Thomas Hunt Morgan made the same connection in his equally hostile *Evolution and Adaptation*, using the term "Darwinian school" to denote the selectionists.[11] Vernon Kellogg's *Darwinism Today* of 1907 was written on the assumption that "Darwinism" now meant the selection theory, with the scientific community being divided into Darwinian and anti-Darwinian camps.[12] The translation of Eberhart Dennert's *At the Deathbed of Darwinism* of 1904 made it clear that it was the

---

[7] George John Romanes, *Darwin and After Darwin*, vol. 2 (new edition, London: Longmans, 1900), pp. 12–13.

[8] Note the subtitle: Alfred Russel Wallace, *Darwinism: An Exposition of the Theory of Natural Selection with Some of Its Applications* (London: Macmillan, 1889).

[9] Samuel Butler, "The Deadlock in Darwinism," reprinted in Butler, *Essays on Life, Art and Science* (London: A. C. Fifield, 1908); on "Wallaceism," see pp. 234–340, esp. p. 236.

[10] William Bateson, *Materials for the Study of Variation: Treated with Especial Regard to Discontinuity in the Origin of Species* (London: Macmillan, 1894), Preface.

[11] Thomas Hunt Morgan, *Evolution and Adaptation* (New York: Macmillan, 1903), Chapters 4–6; see, e.g., p. 462.

[12] Vernon L. Kellogg, *Darwinism Today: A Discussion of Present-Day Scientific Criticism of the Darwinian Selection Theories* (New York: Henry Holt, 1907).

death of the selection theory that was being celebrated.[13] The narrower sense of the term seems to have become increasingly commonplace just as many were becoming aware of how seriously Darwinism (in this narrow sense) was being challenged by scientists. Yet it would have been possible to recognize the strength of anti-Darwinism without necessarily implying that selection had once been the dominant theory. After all, the polarization to which Romanes referred had taken place only in the last decade or so of the nineteenth century. The question now emerges: why would anyone have thought that a rigid selectionism had been the dominant theory (and ideology) during the earlier, more flexible phase of Darwinism?

THE SPECTER OF DOGMATIC DARWINISM

In Britain, at least, this question has to be answered by looking at the resurgence of a movement seeking to consolidate the link between a nonmaterialistic science and a very liberal form of religious belief.[14] The term "resurgence" is important here, because the ideas on which the synthesis was to be based had been clearly articulated during the Victorian era – yet the early twentieth-century exponents of the view went out of their way to suggest that they were creating a new vision that would replace the materialism that they saw as characteristic of the previous century. In order to substantiate this image, they needed to suppress any hint that their ideas had been anticipated during the Victorian era, and the easiest way to do this was to pretend that that era had been one in which materialism had enjoyed practically unchallenged support. Since Darwinism was identified as a key plank in the materialist platform, this meant portraying the Victorians as having been swept away by uncritical enthusiasm for the selection theory. In effect, the Victorians were seen as having been hoodwinked into accepting a scientifically implausible selection theory because it fitted their materialist preconceptions. Teleological and hence non-Darwinian theories of evolution, actually widely developed in the nineteenth century, were introduced as new developments representing a major scientific innovation. This model of the history of evolutionism required the suppression of a number of inconvenient facts. Although there had been few really wholehearted selectionists during the late nineteenth century, the opponents of selectionism had to be depicted as isolated figures martyred by a ruthless Darwinian orthodoxy. The fact that a few genuine Darwinians

[13] E. Dennert, *At the Deathbed of Darwinism*, trans. E. V. O'Harra and John H. Peschges (Burlington, IA: German Literary Board, 1904).
[14] For details of this movement, see Bowler, *Reconciling Science and Religion*.

held strong religious convictions – this would include A. R. Wallace and the Oxford professor of zoology E. B. Poulton[15] – also had to be concealed. And finally, it was necessary to ignore the emergence of a new anti-Darwinian movement that was far more in tune with the way in which early twentieth-century science was developing, but that gave no comfort to the supporters of teleology in evolution: Mendelian genetics.

Despite these inconvenient facts, there were several movements within early twentieth-century British culture eager to present themselves as successfully countering the baleful influence of a dogmatically materialistic Darwinism. Conservatives within the Christian churches, both evangelical and Catholic, were still opposed to any form of evolutionism – at least as applied to the origin of the human race – and it suited them to tar the whole evolutionary movement with the brush of Darwinian atheism. Liberal Christians, the most influential of whom were the Modernists within the Anglican Church, welcomed the general idea of evolution but saw it as essential that the theory allow progress and purpose to be seen at the heart of what was, to them, a divinely instituted process of creation. They too were anxious to distance themselves from the kind of Darwinism associated with the "war" in which T. H. Huxley and other scientific naturalists had assaulted organized religion. Outside the churches, there was a wide circle of vaguely philosophical writers who saw themselves as promoting a new religion in which human activity was the continuation of nature's purposeful striving toward spiritual progress. They also wanted to distance themselves from Darwinian selectionism, and in the case of the playwright Bernard Shaw this tactic included a strong emphasis on the dogmatic character of the Darwinism that, he claimed, had destroyed the credibility of his theoretical predecessor, Samuel Butler.

To all of these writers, it had seemed obvious that there had indeed been a war between science and religion during the Victorian era, and that the materialistic scientists had won a temporary victory that was only now being undermined. To this extent, the strident campaigns of militant scientific naturalists such as T. H. Huxley and John Tyndall had been successful. They may not have converted the majority of scientists to their position – indeed, all the evidence is that they had not – but they created an enduring impression of science as a force opposed to supernaturalism and hence to religious belief. This image seems to have overwhelmed the influence of those scientists who

---

[15] On Poulton, see Richard England, "Natural Selection, Teleology and the Logos: From Darwin to the Oxford Neo-Darwinists, 1859–1909," in J. H. Brooke, M. J. Osler, and J. van der Meer, eds., *Science in Theistic Contexts* (*Osiris*, vol. 11; Chicago: History of Science Society, 2001), pp. 270–87.

still maintained a religious faith, at least in the popular imagination. A survey on *The Religious Beliefs of Scientists* published in 1910 showed that there were many eminent figures who professed some form of belief (although not necessarily orthodox Christianity). Since much of this information had in fact been gathered fourteen years earlier, it should have been obvious that many late Victorian scientists had not been converted to the Huxley–Tyndall axis.[16] Yet such information was usually taken to mean that scientists had at last revolted against the materialists' takeover of their profession.

This distorted view of the Victorian era created an assumption that materialistic Darwinism had dominated biology at the time. Although Huxley himself had never been a dogmatic selectionist, the depiction of nature as a scene of relentless struggle in his "Evolution and Ethics" of 1893 helped to create the impression that he did see evolution on neo-Darwinian terms – but had now come to appreciate how dangerous the extension of this model to human affairs could be.[17] Coupled with the discussions of neo-Darwinism mentioned earlier, Huxley's attack on what would eventually be known as "social Darwinism" helped to create an artifical sense of the degree to which the earlier generation of Darwinists had been committed to the selection theory. The wave of anti-Darwinian theories promoted at the turn of the century was thus seen as a new initiative, not as the continuation of a long-standing Victorian tradition of seeking to reconcile evolutionism with some form of teleology.

Some went even further in their rejection of Darwinism, although Britain did not see a surge of Protestant fundamentalist opposition to evolution like the one that engulfed the United States in the 1920s. There was a small movement along the same lines, with its focus at the Victoria Institute in London, but it never achieved the kind of publicity gained by American advocates of the same position. The American creationist George McCready Price was in London during the early 1920s but struggled to gain an audience for his views. He insisted that the general public had been browbeaten into uncritical acceptance of evolutionism by the activities of materialist scientists.[18] Another American who published in London was Floyd E. Hamilton, whose *The Basis of Evolutionary Faith* argued that the old, confident selectionism of the previous

---

[16] Arthur H. Tabrum, ed., *Religious Beliefs of Scientists: Including One Hundred Hitherto Unpublished Letters on Science and Religion by Eminent Men of Science* (London: Hunter and Longhurst, 1910).

[17] T. H. Huxley, "Evolution and Ethics," in his *Evolution and Ethics and Other Essays* (London: Macmillan, 1894), pp. 46–116. Curiously, Huxley does not use the term "Darwinism" in this essay.

[18] See Ronald L. Numbers, *The Creationists* (New York: Knopf, 1992), Chapter 8.

generation had now broken down, leaving the whole edifice of evolutionism vulnerable.[19] Hamilton insisted that Darwinism was dead, whatever its modern supporters might claim. A similar position had already been maintained in *The Bankruptcy of Evolution* by the Rev. Harold C. Morton, who claimed that the materialistic version of evolutionism had swept the board among an earlier generation of thinkers determined to reject any role for the supernatural.[20] The president of the Victoria Institute from 1927 was the eminent electrical engineer Sir Ambrose Fleming, who wrote a number of books and articles attacking the idea of a natural origin for the human race. He stressed the materialism of the first generation of evolutionists and insisted that Huxley, Haeckel, and others had been so successful that anyone now challenging the Darwinian theory was inviting ridicule.[21]

Throughout the early decades of the century, the most influential opposition to evolutionism came from Roman Catholic writers. Since their numbers included Hilaire Belloc and (after his conversion in 1922) G. K. Chesterton, these writers exerted an influence beyond what might have been expected, given their church's limited membership in Britain. Belloc played havoc with the rationalist view of the history of life and civilization as presented in H. G. Wells' *Outline of History* and was widely supposed – in literary circles, at least – to have won the ensuing debate. Like many others, he proclaimed that "Darwinism is dead," and he ridiculed Wells for failing to realize this.[22] Chesterton took up the same theme, repeating it as late as 1935 in an article for the *Illustrated London News* in which he depicted the original form of Darwinism as a product of the materialism of the Victorian age.[23] The image of Darwinism as a Victorian dogma that no one had dared to challenge was also promoted in Arnold Lunn's *The Flight from Reason* of 1932, published just a year before he joined the Roman Catholic Church.[24]

---

[19] Floyd E. Hamilton, *The Basis of Evolutionary Faith: A Critique of the Theory of Evolution* (London: James Clarke, 1931), p. 18.

[20] Harold C. Morton, *The Bankruptcy of Evolution* (London: Marshall Brothers, 1925), p. 69.

[21] See, for instance, Fleming's "Evolution and Revelation," *Journal of the Transactions of the Victoria Institute* 59 (1927): 11–40.

[22] Hilaire Belloc, *A Companion to Mr. Wells' "Outline of History"* (London: Sheed and Ward, 1926), pp. 10–11. On Belloc's apparent success in the debate with Wells, see, for instance, Michael Corley, *The Invisible Man: The Life and Liberties of H. G. Wells* (New York: Atheneum, 1993), Chapter 7. As noted later, J. B. S. Haldane was provoked into writing in defense of Darwinism in part by the popularity of Belloc's critique.

[23] G. K. Chesterton, "About Darwinism," reprinted in Chesterton, *As I Was Saying: A Book of Essays* (London: Methuen, 1936), pp. 194–9.

[24] Arnold Lunn, *The Flight from Reason: A Criticism of the Dogmas of Popular Science* (London: Eyre and Spottiswood, 1932), p. x.

Among more liberal religious thinkers, the emphasis was more on promoting a progressionist, teleological evolutionism in which the emergence and progress of the human spirit was seen as the culmination of the divine plan of creation. Some writers in this tradition also stressed how their "new" evolutionism transcended the materialism and Darwinism that had been characteristic of the Victorian debates. One of the leading scientific figureheads of the movement was J. Arthur Thomson, the professor of natural history at Aberdeen, who had almost abandoned his studies under Haeckel during the 1880s to train as a minister in the Free Church of Scotland. In his Gifford Lectures for 1915–16, Thomson argued that the tendency of the previous generation of evolutionists to depict the world as a "dismal cockpit" and a "chapter of accidents" had "engendered what may be called a natural irreligion" that he proposed to show was untenable.[25] Like many of those looking for a more teleological vision of evolution, Thomson wanted to minimize the extent to which the process was driven by death and suffering. Although never an outright Lamarckian, he favoured Henri Bergson's vision of "creative evolution" in which the life force struggled toward ever higher forms of expression.

Another Scot, James Young Simpson, who taught biology at the Free Church's New College in Edinburgh, wrote his inappropriately titled *Landmarks in the Struggle between Science and Religion* in an attempt to minimize the impact of the "warfare" metaphor. He insisted that the mechanist worldview of the nineteenth century, including the Darwinian selection theory, was now defunct.[26] His *Nature: Cosmic, Human and Divine* of 1929 continued this theme by stressing that across all the sciences, "Victorian cocksureness and arrogance have been superseded by Georgian hesitancy, or, shall we say, open-mindedness and humility."[27] The alleged power of the Darwinian theory was again seen as a direct product of its ability to chime with an all-powerful ideology of materialism that had dominated Victorian science and thought.

The most powerful advocate of this viewpoint was George Bernard Shaw, who was promoting his own vision of "creative evolution" as a rival both to materialism and to orthodox Christianity. Shaw's vitriolic critique of Darwinian

---

[25] J. Arthur Thomson, *The System of Animate Nature: The Gifford Lectures Delivered in the University of St. Andrews in the Years 1915 and 1916* (London: Williams and Norgate, 1920), vol. 1, p. v.

[26] James Young Simpson, *Landmarks in the Strugggle between Science and Religion* (London: Hodder and Stoughton, 1925), p. 213.

[27] James Young Simpson, *Nature: Cosmic, Human and Divine* (New Haven, CT: Yale University Press, 1929), p. 6.

materialism in the Preface to his *Back to Methuselah* of 1921 is so well known that it need not be repeated here. But his perception of the theory's status as a Victorian dogma is very clear, and seems to have arisen largely because he was aware that his views on the need for life to be seen as a creative force had been anticipated by Samuel Butler. Shaw knew that Butler had been ostracized by the Darwinian community in the years shortly before Darwin's death and insisted that this was because Butler's Lamarckian views contradicted the materialistic dogma of the Darwinians. He claimed that at that time it had been impossible to criticize Darwinism without reproof, so strong was the Darwinians' hold on scientific opinion. The Victorians had been so glad to escape from the old idea of a static Creation that they failed to realize that the alternative offered to them by Darwin was even more frightening because it eliminated any role for mind in the workings of nature.[28] Darwin had been able to "convert the crowd" because his theory was so simple to explain, while the idea that evolution depended on the will power of organisms seemed mystical. Shaw realized that Butler had been dismissed so brusquely because he had been impolitic enough to be rude to Darwin himself, but he also seems to have felt that Butler would have been ridiculed even had this personal element not been introduced into the situation. Here, then, was one of the most vocal advocates of a more teleological evolutionism presenting Victorian Darwinism as a dogma that it had been impossible to challenge without being dismissed as a crank.

Shaw's Lamarckism was, by then, largely out of date as far as most scientists were concerned (but see Chapter 8 of this volume). Even those who still favored some form of teleological evolutionism had to express their hopes in a form that was less open to refutation by the increasingly dominant Mendelians. But his comments about Butler remind us that there had never been a coherent school of neo-Lamarckism in Britain – in part because Butler's personal animosity toward Darwin had made it impossible for him, or anyone identified with him, to serve as a figurehead. The fact that prominent "Darwinians" such as Herbert Spencer had actively supported the inheritance of acquired characteristics was conveniently forgotten. They could not be "Lamarckians," because they were already identified with the opposing school. Shaw was evidently unaware of the strong American tradition of neo-Lamarckism – but so, apparently, were most of his British readers. They knew that there had been no coherent neo-Lamarckian school on their side of the Altantic, and this was enough to substantiate the claim that Darwinism had been triumphant. It is

---

[28] George Bernard Shaw, *Back to Methuselah: A Metabiological Pentateuch* (London: Constable, 1921), pp. xlii–xlviii.

possible that the rhetorical ineptness of anti-selectionists such as Butler contributed to the growing impression that the whole scientific community had
endorsed the selection theory. His personal antagonism to Darwin ensured
that Butler would be ostracized, even though there were many biologists who
shared some of his reservations about the adequacy of the selection theory.
The politics of the Victorian scientific community thus paved the way for a
later generation to misrepresent the extent to which that community had been
committed to a materialistic Darwinism.

THE "NEW" NATURAL THEOLOGY

Those who supported the "new" reconciliation between science and religion
based their optimism on the hope that science had turned its back on what
they perceived to have been the dogma of Victorian materialism. A new natural theology would be created that would take evolution on board by treating
it as a divinely instituted process of development aimed at the production of
spiritual beings and at the progress of those beings toward some future state
of perfection. In fact, of course, the Victorian churches had gone a long way
toward reconciling their faith with evolutionism after the short-term hostility
manifested in episodes such as the debate between T. H. Huxley and Samuel
Wilberforce (which became a key element in the articulation of the "warfare" mythology). The 1889 volume *Lux Mundi*, edited by the Anglo-Catholic
Charles Gore, had gone a long way toward showing how even relatively conservative Christians could accept "creation" as a metaphor compatible with a
vaguely naturalistic evolutionism. In fact, evolutionism had to retain a role for
teleology in order for this move to work, and the new natural theology (actually created in the late nineteenth century) tended to prefer Lamarckian and
other non-Darwinian approaches that allowed for either the creative activity
of individual organisms or some predetermined progressive trend aimed at the
production of mind and spirit. Without apparently recognizing the pedigree
of this way of thinking, the theologians and philosophers of the early twentieth
century saw themselves as participating in a new initiative to reconcile science
and religion.

   An early example of the excitement that could be created by this liberal
compromise can be seen in the reaction to the "New Theology" promoted by
the Congregationalist minister R. J. Campbell in 1907.[29] Campbell called in

---

[29] Reginald John Campbell, *The New Theology* (London: Chapman and Hall, 1907); on the
   resulting controversy, see Keith W. Clements, *Lovers of Discord: Twentieth-Century Theological
   Controversies in England* (London: Society for the Promotion of Christian Knowledge, 1988),
   Chapter 2, and Bowler, *Reconciling Science and Religion*, Chapter 7.

Oliver Lodge and Bernard Shaw to speak at his meetings – the latter being an extremely dangerous move, given that Shaw openly proclaimed his version of creative evolution as an alternative to Christianity. God, in Campbell's view, was seen as immanent within the world rather than as transcendent; evolution was His way of becoming self-conscious within that world; and original sin was dismissed as no more than a necessary relic of the process of creation, to be overcome by following the perfect man, Christ. More conservative Free Church ministers pointed out the dangers of so openly flouting the traditional emphasis on sin and the need for redemption. Campbell eventually repudiated the New Theology and was received into the Anglican Chuch by Gore. In effect, his appeal to Shaw had pinpointed the problem with the New Theology – by the standards that had been accepted for centuries, it simply wasn't Christianity, because it left no room for Christ as the Saviour and found it difficult to see any point in his agony on the Cross.

The most active group of liberals were the Modernists within the Anglican Church, now well organized with a society, the Modern Churchmen's Union, and a journal, the *Modern Churchman*. The Union ran annual conferences, several of which had themes related to the new natural theology and included sympathetic scientists among the speakers.[30] The Modernists were not, of course, modernists in the same sense as the artistic and literary avantgarde of the early twentieth century: indeed, their plea for a reconciliation of Christianity with "modern" thought was really aimed at catching up with the worldview of the late nineteenth century. Many of them were philosophical idealists, reflecting the viewpoint that had dominated academic philosophy at the time of their training in the last decades of the nineteenth century. They were thus primed to take on board a nonmaterialistic intepretation of science that would include the revival of vitalist physiology and the teleological evolutionism favoured by the Lamarckians and related opponents of Darwinism.

To put a little more flesh on the bones of this episode, I will briefly outline the views of two figures who received wide public attention but whose interpretations reveal the extent of the communication problem between the Modernists and the more active currents of scientific research during the 1920s and 1930s. Charles Raven, canon of Liverpool and later a professor of divinity at Cambridge, is best known to historians of science for his

---

[30] See Alan M. G. Stephenson, *The Rise and Decline of English Modernism* (London: Society for the Promotion of Christian Knowledge, 1984), which gives details of the annual conferences in an appendix; see also Clements, *Lovers of Discord*, Chapters 4 and 5, and Bowler, *Reconciling Science and Religion*, Chapter 8.

biography of the seventeenth-century naturalist John Ray. But Raven was an active Modernist who genuinely believed that the program of liberalization would reinvigorate Christianity. His work as a historian of science was in part undertaken to argue that the harmony of science and religion displayed by the earlier natural theologians needed to be reestablished if materialism was not to destroy Western civilization. He was the driving force behind a church congress in 1926 that sought to promote the Modernist emphasis on a God Who is immanent within nature, and in the following year his book *The Creator Spirit* outlined the case for a nonmaterialistic biology as the foundation for a renewed natural theology.[31] Raven had studied briefly under William Bateson and had conceived a hatred for the concept of genetic determinism. He was an open supporter of Lamarckism, praising the work of the pathologist J. George Adami in this area and bringing him in to speak at the 1926 congress.[32] He was most enthusiastic about the views of Lloyd Morgan, whose philosophy of emergent evolution chimed with his own efforts to show that the creative behavior of animals could influence evolution. He never wavered in his contempt for both genetic determinism and Darwinism, and in later years became an enthusiastic supporter of the theistic evolutionism of Teilhard de Chardin.[33] Raven provides us with a perfect example of the Modernist clergyman, convinced by the writers of a previous generation that Darwinism was dead and unable to comprehend that it was really undergoing a renaissance.

Few Modernist writers showed any more appreciation than Raven of what was actually going on in scientific evolutionism. An important exception – and for a very special reason – is my other example, Ernest William Barnes, who began his career as an applied mathematics teacher at Cambridge, becoming a Fellow of the Royal Society for his research before becoming ordained and being appointed canon of Westminster and, in 1924, bishop of Birmingham. Barnes was notorious for delivering what the press called his "gorilla sermons,"

[31] Charles E. Raven, *The Eternal Spirit: An Account of the Church Congress Held at Southport, October 1926* (London: Hodder and Stoughton, 1926); Raven, *The Creator Spirit: A Survey of Christian Doctrine in the Light of Biology, Psychology and Mysticism* (London: Martin Hopkinson, 1927); Raven, *John Ray, Naturalist: His Life and Work* (Cambridge: Cambridge University Presss, 1942). On Raven's life, see F. W. Dilliston, *Charles Raven: Naturalist, Historian, Theologian* (London: Hodder and Stoughton, 1975).
[32] J. George Adami, "The Eternal Spirit of Nature as Seen by a Student of Science," *Modern Churchman* 19 (1926): 509–19; for Adami's Lamarckism, see his *Medical Contributions to the Study of Evolution* (London: Duckworth, 1918).
[33] Charles E. Raven, *Natural Religion and Christian Theology*, 2 vols. (Cambridge: Cambridge University Press, 1953); Raven, *Teilhard De Chardin: Scientist and Seer* (London: Collins, 1962).

in which he pointed out the need for the church to be honest in admitting how much of its traditional dogma would have to be abandoned if evolution theory were to be accepted.[34] Even a progressionist, teleological evolutionism required a reinterpretation of the doctrine of original sin. To begin with, Barnes said little about the actual process of evolution, but he seems to have assumed that it was purposeful and aimed at the production of higher mental states. In 1930, though, he obtained a copy of R. A. Fisher's *Genetical Theory of Natural Selection* and began a correspondence with Fisher, who had studied under him while a student at Cambridge. Barnes was one of the few clergymen who could actually understand Fisher's mathematics (although even he admitted it was hard going) and from this point on he made a point of acknowledging the growing influence of the Darwinian theory. He did not concede that the selection theory offered a complete explanation, and he continued to believe that evolution was intended to produce beings with higher mental and spiritual qualities, but he was now aware that the more simpleminded forms of teleology were unacceptable.[35] Fisher himself was a practicing Anglican, and he went on to argue that natural selection was itself a creative process which made Bergson's *élan vital* unnecessary – but he did not publish on this until much later.[36]

If more clergymen had enjoyed Barnes's opportunities to collaborate with someone engaged in cutting-edge research in evolutionary theory, the wave of enthusiasm for the new natural theology might have been undermined. Given the specter of Darwinian materialism created by the earlier body of literature, few would have been willing to endorse Fisher's claim that the selection mechanism was, after all, a suitable vehicle for the divinely ordained process of creation. In fact, such a confrontation did not arise, because the climate of opinion within theology was changing even more quickly than the climate in science. The confident progressionism of the Modernists seems to have attracted a good deal of support within the church during the 1920s, but in the increasingly pessimistic atmosphere of the following decade, appeals to the perfectibility of humankind as the culmination of a divine plan seemed increasingly unrealistic. Modernism fell from favor as many clergymen turned to a neo-orthodoxy modeled on the theology of Karl Barth, in which any

[34] See John Barnes, *Ahead of His Age: Bishop Barnes of Birmingham* (London: Collins, 1979).

[35] See the chapters on biology in John Barnes, *Scientific Theory and Religion: The World Described by Science and Its Spiritual Intepretation* (Cambridge: Cambridge University Press, 1933). Much of this book is devoted to a highly technical analysis of physical and cosmological theory, where Barnes's training made him an expert.

[36] R. A. Fisher, *Creative Aspects of Natural Law* (Cambridge: Cambridge University Press, 1950).

form of natural theology was seen as irrelevant to the need for salvation of a humanity totally alienated from God by sin.

The Modernist literature made frequent reference to the popular writings of scientists, and it seems evident that it was important to these theologians and philosophers that they be able to substantiate the claim that science had turned its back on materialism. By quoting eminent figures within the scientific community who openly endorsed their antimaterialist position, they were able to create the impression that the tide had indeed turned against Darwinism. There was an implied assumption that the scientific community as a whole had changed tack, with the remaining scientific rationalists and materialists themselves becoming isolated and marginalized. From the perspective of hindsight, the reaction against materialism among working biologists was much less general and much more short-lived than its theological and philosophical supporters would have wished. The new developments of the 1920s and 1930s indicated that the scientific community was beginning to recognize that a materialistic and Darwinian perspective was the best way forward, whatever the doubts expressed at the turn of the century. Yet the rationalist scientists of the period seem to have had only limited success in getting this message across. In 1930, when all the evidence shows that it was just coming into its own within scientific biology, a significant proportion of nonspecialist writers were still convinced that Darwinism was dead. This was because a few senior figures were able to exploit their reputations and their access to the means of publication to play a disproportionate role in shaping the public understanding of science. These figures were either indifferent to or unaware of the revival of Darwinism – yet their books and articles were the ones that were read by theologians and many other nonspecialists, and the image of science that they created still shaped the popular image of science as a whole.

The names of these scientists crop up with monotonous regularity in the writings of liberal theologians and antimaterialist philosophers. They wrote prolifically and were skilled at presenting their material to a lay readership; indeed, most of them virtually abandoned scientific research for careers as writers, educators, and administrators. They were active in contributing to a number of collected volumes designed to highlight the involvement of scientists in the "new" natural theology.[37] Here we can do no more than

---

[37] See, e.g., Frances Mason, ed., *Creation by Evolution: A Consensus of Present-Day Knowledge as Set Forth by Leading Authorities in Non-Technical Language that All May Understand* (New York: Macmillan, 1928); and Mason, ed., *The Great Design: Order and Progress in Nature* (London: Duckworth, 1934).

list their names and hint at the range of their activities, but the success of their enterprise must be judged by the frequency with which their work was cited by the nonscientific writers who also participated in the campaign against materialism. Several played active roles in conferences and meetings organized by religious bodies, as when Oliver Lodge spoke in support of Campbell's New Theology. The 1925 conference of the Modern Churchmen's Union on The Scientific Approach to Religion was addressed by Lloyd Morgan, while Lodge and J. A. Thomson spoke at the 1931 conference on Man.[38]

The psychologist William McDougall actively promoted the idea that the mind existed independently of the material body.[39] He also performed an experiment that he claimed offered support for the Lamarckian theory, and it is clear that the advocates of teleological evolutionism had a strong preference for this theory, although they were aware that there it was encountering growing hostility among experimental biologists. But the two evolutionists most frequently cited in support of the new natural theology were J. Arthur Thomson and Conwy Lloyd Morgan. As noted earlier, Thomson had at first been an enthusiatic Christian, but he was soon converted to a more generalized theism by the influence of the sociologist and town planner Patrick Geddes, with whom he subsequently collaborated on a number of publishing projects.[40] Thomson was appointed professor of natural history at Aberdeen, and he seems gradually to have abandoned research for a career devoted to teaching and writing. Along with Geddes, he published textbooks intended to allow biology to be taught in a nonmechanistic fashion.[41] He also published an impressive array of books and articles arguing directly for the new natural theology and a vision of evolution as a creative process driven by the power of the mind. These include his Gifford Lectures, published as *The System of Animate Nature*, along with *Science and Religion, Purpose in Evolution*, and *The Gospel of Evolution*.[42]

---

[38] See Stephenson, *The Rise and Decline of English Modernism*, Appendix A.

[39] William McDougall, *Body and Mind: A History and Defence of Animism* (London: Methuen, 1911); McDougall, *Modern Materialism and Emergent Evolution* (London: Methuen, 1929); McDougall, *The Riddle of Life: A Survey of Theories* (London: Methuen, 1938).

[40] See Thomson's letters to Geddes in the Geddes papers, National Library of Scotland.

[41] See, e.g., J. A. Thomson and Patrick Geddes, *Life: Outlines of General Biology*, 2 vols. (London: Williams and Norgate, 1931).

[42] Thomson's *The System of Animate Nature* has already been cited for its stated purpose of destroying Victorian materialism; see also J. A. Thomson, *Science and Religion* (London: Methuen, 1925); Thompson, *Purpose in Evolution* (Oxford: Oxford University Press, 1932); Thompson, *The Gospel of Evolution* (London: George Newnes, n.d.)

Lloyd Morgan had made his name as an evolutionary psychologist during the 1890s and had been one of the codiscoverers of the mechanism of
"organic selection" (also called the "Baldwin effect," after J. M. Baldwin), in
which natural selection adapts the species to new behavior patterns chosen by
the organism.[43] He was appointed professor of zoology at Bristol and eventually became vice chancellor of the university. His later work continued to stress
how the mental activities of animals had a real effect on their lives and on the
course of evolution, in a way that could not be explained on the basis of purely
materialistic principles. He had always insisted that the higher human mental
faculties should not be attributed to animals – this is the essence of "Lloyd
Morgan's canon" in animal psychology. The human mind stands above the
animal mind, just as the animal mind transcends material nature. This sense
of new levels of activity emerging in the course of evolution became the central
feature of his Gifford Lectures for 1922–23, published as *Emergent Evolution*
and *Life, Mind and Spirit*.[44] Morgan made it clear that for him, evolution was
intended to produce these higher levels by the Creator: the detailed course
of evolution might not be predetermined, but its major stages of development were, in the sense that each upward step was bound to occur sooner or
later.

The only other scientist to be cited as frequently as Thomson and Morgan
in favor of teleological evolutionism was the physicist Oliver Lodge. Lodge
made his name in the study of radiation during the late nineteenth century
and was a leading advocate of the theory of the ether, which he conceived
as the basis for an alternative to scientific naturalism because it allowed the
cosmos to be seen as a coherent whole. But Lodge was also a leading suporter
of spiritualism, and he linked his two main interests by arguing that evolution
was a purposeful process designed to create the human spirit – which would
then undergo its own ethical progress in the next world.[45]

---

[43] See Robert J. Richards, *Darwin and the Emergence of Evolutionary Theories of Mind and Behavior*
(Chicago: University of Chicago Press, 1987), Chapter 8.

[44] Lloyd Morgan, *Emergent Evolution* (London: Williams and Norgate, 1923); Morgan, *Life, Mind
and Spirit* (London: Williams and Norgate, 1926).

[45] Lodge wrote endlessly on this theme; among his many books are *Man and the Universe*
(London: Methuen, 1908), *Raymond: or Life and Death* (London: Methuen, 1916), *The Making
of Man: A Study in Evolution* (London: People's Library, 1929), and *Evolution and Creation*
(London: Hodder and Stoughton, 1926). On Lodge's ether physics and its wider implications,
see Bryan Wynne, "Physics and Psychics: Science, Symbolic Action and Social Control in Late
Victorian England," in Barry Barnes and Steven Shapin, eds., *Natural Order: Historical Studies
of Scientific Culture* (Beverley Hills, CA: Sage, 1979), pp. 167–87.

Between them, these senior figures published enough nonspecialist literature to convince at least those readers predisposed to favor the new natural theology that a nonmaterialist and non-Darwinian trend was firmly established in science. But was this a correct assessment of the situation by the 1920s? Whatever the developments in physics, hindsight suggests that in fact the wave of opposition to materialism in biology was relatively limited in both influence and duration, so that by the second and third decades of the century the majority of working scientists would no longer have endorsed such views. This was certainly the opinion of the rationalists, desperately striving to stem the apparent tide of enthusiasm for the new natural theology. The veteran rationalist campaigner Joseph McCabe reissued his *The Existence of God* in 1933 to proclaim that the scientists supporting the antimaterialist position were all out of date – they were "a lingering group of elderly men . . . whose watches stopped forty years ago."[46] There was some truth to McCabe's argument: the figures we have mentioned were all born before 1875; by the 1920s they were nearing retirement and, in some cases, death. Outside Britain, holistic and organismic views retained some influence in biology and were by no means limited to those scientists with strong religious beliefs. W. M. Wheeler's work in America illustrates this point, as does Anne Harrington's account of German science in the period.[47] But in Britain, biologists such as Joseph Needham who advocated organicist views seem to have found it hard to gain a hearing. In this respect, McCabe's analysis accurately reflected the direction of scientific thinking in Britain when it depicted Morgan's emergent evolutionism as more characteristic of the previous generation's thinking.

The flood of literature in support of the new natural theology thus dried up as the scientists involved became incapacitated or died. But for a couple of decades, at least, their activities had a sufficiently high profile to convince a generation of readers that the collapse of Darwinism had paved the way for a reconciliation between religion and science. Of course, the opposing position was being promoted by rationalists such as H. G. Wells, E. Ray Lankester, and Arthur Keith. But there was obviously heavy resistance to this in some circles – literary people seem to have felt that Belloc got the better of Wells in the debate over the *Outline of History*, and we have seen that many theologians also welcomed the anti-Darwinian rhetoric. It has been suggested that

---

[46] Joseph McCabe, *The Existence of God*, rev. ed. (London: Watts, 1933), pp. 142–3.
[47] Chapter 8, this volume; Anne Harrington, *Reenchanted Science: Holism in German Culture from Wilhelm II to Hitler* (Princeton, NJ: Princeton University Press, 1996).

J. B. S. Haldane wrote his 1932 survey *The Causes of Evolution* as a response to Belloc's critique.[48] The anti-Darwinian rhetoric of the previous decades had clearly had a significant effect in blinding many ordinary people to the fact that the selection theory was now at last coming into its own. Popularizers such as Haldane and Huxley would have to work hard to convince the reading public that the specter of Darwinism was about to haunt them again.

[48] See Gordon McOuat and Mary P. Winsor, "J. B. S. Haldane's Darwinism in Its Religious Context," *British Journal for the History of Science* 28 (1995): 227–31. Haldane became a Marxist at about this time, but that is another story.

CHAPTER FOUR

Natural Atheology

Abigail Lustig

Evolutionary biologists, especially in the United States, seem to be engaged in a perpetual war with religion. On the face of it, this is unsurprising: over two-thirds of the American population belong to religious congregations; nearly half describe themselves as born again into evangelical Christian faiths that depend on revelation and the doctrine of justification by faith; and a sizeable and vocal proportion of these consider the teaching of Darwinian evolutionary biology to be anathema. All U.S. (and to a lesser extent, British) evolutionists must therefore choose sides in an ongoing cultural and political conflict. But the public evangelists for evolution, by and large, do more than defend the validity of their science; they also carry the war into the enemy's camp, aiming not only to safeguard their own work but also to vitiate the very underpinnings of religion.

It is generally seen as unfortunate when scientists let their religious or other metaphysical beliefs inform their science; the philosopher of evolution Michael Ruse, for example, speaks disapprovingly in his *Mystery of Mysteries: Is Evolution a Social Construction?* of "cultural values built right into [Julian Huxley's] science" and of "cultural-value infiltration" into the work of the architects of the Modern Synthesis and current evolutionary biologists – including the late Stephen Jay Gould, Richard Dawkins, Richard Lewontin, and E. O. Wilson – as though evolutionary biology subscribed at least as much as any other science to the "metavalue . . . of the internal culture of science itself, namely, that of keeping science distinct from culture and hence nonepistemic-value-free."[1] But evolutionary biology, perhaps more than any other science, not only is not nonepistemic-value-free, but, by virtue of its descent, cannot be so. Born in theology, its goals entail the extension of an a priori metaphysical rationalism whose aim at its origin was to upset the strongest rational argument for the existence of God.

Because it is an historical science, evolutionary biology, more than any other, is dependent upon the words that construct its theories. Its aim is to elucidate how history happens, and its fundamental tenet, drawn from the principle of natural selection, is that the actual future is unpredictable within general limits imposed by the past, by physics, and by the biological constraints on variation. Its theories are judged, like those of other historical disciplines, on the basis of relative plausibility; every explanatory narrative is created as an answer to and transformation of previous ones. Modern evolutionary biology traces its descent (with modifications) from Charles Darwin, and most particularly from the *Origin of Species* of 1859. Darwin's work has a continuing textual and rhetorical importance for his successors that is unparalleled elsewhere in science (with the dubious exception of Freudian psychology). The *Origin*, a grand piece of historiography made up of many smaller illustrative narratives,[2] was itself created as a response to one of the great conundrums of natural history – the order and diversity of life – and to one of its most convincing answers – the theological argument from design.

Darwin's own text owes its greatest debt, in narrative and overall rhetorical strategy, to William Paley's *Natural Theology* of 1802 – the acme of a natural theology literature dating back to the seventeenth century, which Darwin read while at Cambridge and, he said in his autobiography, virtually got by heart. Paley made what is still the best and clearest statement of the argument from design: that as the existence of the watch or the telescope – clearly artifices constructed for particular purposes – demonstrates the existence of a watch and telescope designer, so the existence of the eye, or the wing, or social instincts – artifices likewise clearly constructed for their various purposes and superior to any human contrivance – demonstrates the existence of a

---

[1] Michael Ruse, *Mystery of Mysteries: Is Evolution a Social Construction?* (Cambridge, MA: Harvard University Press, 1999), pp. 98, 188, 117.

[2] On the subject of evolutionary biology as historiography, see also G. Beer, *Darwin's Plots: Evolutionary Narrative in Darwin, George Eliot, and Nineteenth-Century Fiction* (London: Routledge and Kegan Paul, 1983); R. J. Richards, "The Structure of Narrative Explanation in History and Biology," in M. Nitecki, ed., *History and Evolution* (Albany: State University of New York Press, 1992); A. J. Lustig, "George Eliot, Charles Darwin and the Labyrinth of History," *Endeavour* 23 (1999): 110–13.

Designer for them as well.[3] Paley's elaboration of the argument from design comprehended three strategic arguments that Darwin would turn to his own ends and that, through Darwin, continue to pervade evolutionary biology today.

First among these is the one that made for Paley's strongest case and one of Darwin's weakest: the perfection of many contrivances in nature. Paley's premise, of course, is that teleology is embedded in all adaptation; creation and purpose are indistinguishable, and to deny the reality of such "evidences of art and skill" is both "absurdity" and "atheism." Paley devotes the whole of Chapter 3 to an analysis of the perfection of the eye as an adaptation for vision and the subtlety of the variations it undergoes according to the needs of various animals. He reiterates the argument of his opening passage, substituting the telescope for the watch and comparing it to the eye point by point; on every point, the eye is, when not equivalent, superior. Finally, Paley confronts the great difficulty of this argument: why should an omnipotent Creator have fussed about with the laws of optics and physics at all? Why should a Deity be so constrained? The eye's very perfection calls the existence of God into doubt:

One question may possibly have dwelt in the reader's mind during the perusal of these observations, namely, Why should not the Deity have given to the animal the faculty of vision at once? . . . Why resort to contrivance, where power is omnipotent? Contrivance, by its very definition and nature, is the refuge of imperfection.[4]

---

[3] "In crossing a heath, suppose I pitched my foot against a *stone*, and were asked how this stone came to be here; I might possibly answer, that for anything I knew to the contrary, it had lain there forever: nor would it perhaps be very easy to show the absurdity of this answer. But suppose I had found a *watch* upon the ground, and it should be inquired how the watch happened to be in that place; I should hardly think of the answer which I had before given, that, for anything I knew, the watch might have always been there. Yet why should not this answer serve for the watch as well as for the stone? Why is it not as admissible in the second case, as in the first? For this reason, and for no other, viz. that, when we come to inspect the watch, we perceive (what we could not discover in the stone) that its several parts are framed and put together for a purpose. . . . This mechanism being observed (it requires indeed an examination of the instrument, and perhaps some previous knowledge of the subject, to perceive and understand it; but being once, as we have said, observed and understood,) the inference, we think, is inevitable; that the watch must have had a maker; that there must have existed, at some time, and at some place or other, an artificer or artificers, who formed it for the purpose which we find it actually to answer; who comprehended its construction, and designed its use." William Paley, *Natural Theology; or, Evidences of the Existence and Attributes of the Deity. Collected from the Appearances of Nature* (London: Printed for R. Faulder, 1802), Chapter 1.

[4] Paley, *Natural Theology*, Chapter 3.

Paley finds his answer in the rational adherence by the Creator to the general laws He had laid down for the universe; without this rationality, we should not be able to deduce His existence:

The question is . . . of very wide extent; and amongst other answers which may be given to it, beside reasons of which probably we are ignorant, one answer is this: It is only by the display of contrivance, that the existence, the agency, the wisdom of the Deity, could be testified to his rational creatures. . . . Take away this, and you take away from us every subject of observation, and ground of reasoning. . . . Whatever is done God could have done without the intervention of instruments or means: but it is in the construction of instruments, in the choice and adaptation of means, that a creative intelligence is seen. . . . God, therefore, has been pleased to prescribe limits to his own power, and to work his ends within those limits.[5]

The adherence of Creation to regular laws that entail the necessity for optimal contrivance are thus for Paley simultaneously a test and a demonstration of God's existence: a test that is posed to "his rational creatures" to unravel and reveal Him, and a demonstration of His own fidelity to natural law.

The use to which Darwin put this chapter of Paley is instructive in understanding the inside-out transformation of natural theology into evolutionary biology. Darwin took up the eye as the centerpiece of the *Origin*'s chapter on "Difficulties on Theory." After discussing a spectrum of eyes, from simple to complex, in the *Articulata*, Darwin meets Paley's challenge head-on:

It is scarcely possible to avoid comparing the eye to a telescope. We know that this instrument has been perfected by the long-continued efforts of the highest human intellects; and we naturally infer that the eye has been formed by a somewhat analogous process.[6]

Darwin, however, inverts both Paley's premises and his conclusions. Why would – not *should*, but *would* – God create as man does, subjected to the laws of physics? It may at first sight seem logical to accept Paley's argument from design, that the eye is designed just as the telescope is. But it is not. Rather, the rational creature, whose "reason ought to conquer his imagination," will find a more plausible, and less complacent, explanation in the operation of solely natural forces:

But may not this inference be presumptuous? Have we any right to assume that the Creator works by intellectual powers like those of man? If we must compare the eye to an optical instrument, we ought in imagination to take a thick layer of transparent

---

[5] Ibid.
[6] Charles Darwin, *On the Origin of Species by Means of Natural Selection, or, The Preservation of Favoured Races in the Struggle for Life* (London: John Murray, 1859; reprinted Cambridge, MA: Harvard University Press, 1964), p. 188.

tissue, with a nerve sensitive to light beneath,... continually changing slowly in density.... Further we must suppose that there is a power always intently watching each slight accidental alteration . . . ; and carefully selecting each alteration which . . . may in any way, or in any degree, tend to produce a distincter image.... Let this process go on for millions on millions of years . . . on millions of individuals; . . . and may we not believe that a living optical instrument might thus be formed as superior to one of glass, as the works of the Creator are to those of man?[7]

Who or what the "Creator" is here is left artfully, disingenuously vague; the term either stands directly for the process of natural selection itself or refers to an ultimate Creator who has done nothing, directly, to create the eye. This latter would be not only a God who, like Paley's, binds himself to consistency with the laws of physics, but also one whose creation with regard to all life (and, by implication, to man) is confined at best to, as Darwin later says, that "one primordial form, into which life was first breathed."[8]

The second strategy Darwin took from Paley was the explanation of the universe's imperfections. Here, however, the positions of relative strength are reversed: where perfection was Paley's Q.E.D. and imperfection required explaining away, for Darwin it was just the opposite. Darwin gloried in imperfection of form, because no rational God, apparently, would include appendixes: "the same reasoning power which tells us plainly that most parts and organs are exquisitely adapted for certain purposes, tells us with equal plainness that these rudimentary or atrophied organs, are imperfect and useless."[9]

The cruelty and profligacy that occur so often in nature Paley had had to explain away (with a strategem that Darwin was often to employ himself to ex-plain away inconvenient facts such as the geological record) with the assertion that "from the confessed and felt imperfection of our knowledge, we ought to presume, that there may be consequences of [the natural] economy which are hidden from us."[10] Paley had well understood the principle of Malthusian overpopulation; he construed the twin ills of predation and superfecundity as compensatory one for the other, and admitted that they presented a diffi-culty to his theory: "Animal properties . . . which fall under this description, do not strictly prove the goodness of God: . . . forasmuch as they must have been found in any creation which was capable of continuance, although it is possible to suppose, that such a creation might have been produced by a being, whose views rested upon misery."[11] Paley's only compensation was that

---

[7] Darwin, *Origin*, pp. 188–9.
[8] Ibid., p. 484.
[9] Ibid., p. 453.
[10] Paley, *Natural Theology*, Chapter 26.
[11] Ibid.

fecundity increased the amount of potential happiness in the cosmos:

It is a happy world after all. The air, the earth, the water, teem with delighted existence. In a spring noon, or a summer evening, on whichever side I turn my eyes, myriads of happy beings crowd upon my view. . . . Swarms of new-born *flies* are trying their pinions in the air. Their sportive motions, their wanton mazes, their gratuitous activity, their continual change of place without use or purpose, testify their joy, and the exultation which they feel in their lately discovered faculties.[12]

Darwin, however, was able to construe predation and superfecundity as the natural outcomes of the struggle for existence, which has profligacy as its very premise. Again, in a passage on the suffering underpinning animal existence, he directly inverted Paley's rhetoric:

We behold the face of nature bright with gladness, we often see superabundance of food; we do not see, or we forget, that the birds which are idly singing round us mostly live on insects or seeds, and are thus constantly destroying life; or we forget how largely these songsters, or their eggs, or their nestlings, are destroyed by birds and beasts of prey. . . .[13]

Again, Darwin sets an implicit test for God, one that Paley's God had passed, if barely, in granting happiness in such great measure to his Creation, but one that Darwin calls into severe doubt.

The third plank underlying Darwin's argument for natural selection is the argument from homology: that the likeness of the appendages of various arthropods, the similarity in leaves and the parts of flowers in flowering plants, and, most famously, the essential similarity of vertebrate skeletons are most parsimoniously explained by the supposition of common descent. This too is the transformation of a natural theological argument spelled out by Paley (and further developed by the comparative anatomist Richard Owen). Paley took the general principle of homology as a central demonstration of the logic of the argument from design: "Whenever we find a general plan pursued, yet with such variations in it as are, in each case required by the particular exigency of the subject to which it is applied, we possess . . . the strongest evidence that can be afforded of intelligence and design; an evidence which most completely excludes every other hypothesis."[14]

---

[12] Ibid.

[13] Darwin, *Origin*, p. 62. Darwin also echoes Paley in the rather abrupt and incongruous conclusion to the otherwise bleak chapter on the "Struggle for Existence": "When we reflect on this struggle, we may console ourselves with the full belief, that the war of nature is not incessant, that no fear is felt, that death is generally prompt, and that the vigorous, the healthy, and the happy survive and multiply" (p. 79).

[14] Paley, *Natural Theology*, Chapter 12.

Darwin, characteristically, inverts the logic while leaving the argument intact. Parsimony and reason demand that similarity derive from common origin, not in an original plan of Creation but in common descent. In an implicit version of the perfection argument, he asks why an omnipotent Creator should constrain Himself to the reuse of a few forms, when even the laws of physics would permit a near-infinity. "How inexplicable are these facts on the ordinary view of creation! ... On the theory of natural selection, we can satisfactorily answer these questions."[15] The argument, however, continues to rely for its strength on what Darwin calls "the ordinary view of Creation" – in other words, on a series of assumptions about what God would or would not do. The parsimony, and therefore the reasonableness, of the explanation by common descent rests upon the reader's judgement of Darwin's own implicitly atheological view vis-à-vis Paley's theological one.

The "one long argument" of the *Origin of Species* thus recapitulates both the structure and the content of Paley's statement of the argument from design. Darwin, of course, turns the argument inside out: all of his rhetoric, all of his narratives, are designed to demonstrate the *nonexistence*, or at least the *nonnecessity*, of God as a proximate cause of the historical development of living things. Nevertheless, the framework of the argument that Darwin adopted from Paley virtually enforced the personification of the disembodied force that he saw as replacing Paley's God – natural selection – and consequently enforced also its teleological emphasis.[16]

Modern evolutionary biologists who have been interested in the implications of evolutionism beyond the reconstruction of the history of life have continued to use these same strategies that Darwin derived from Paley as the implicit framework of their explanatory narratives. This skeleton of common purpose, this homology of argument, mean that an essentially theological – because *a*theological – metaphysics involving an a priori commitment to scientistic explanation and enforcing a standard of judgment for the plausibility of (evolutionary) explanations has been implicit in evolutionary

---

[15] Darwin, *Origin*, p. 437. Homology actually has two aspects, both of which Darwin took as demonstration of descent with modification: homology among classes (vertebrate skeletons) and within the individual (the homology of bones of the skull with vertebrae).

[16] John F. Cornell in 1987 analyzed some of the theological influences on Darwin's own formation of the idea of natural selection in terms of deism and its associated concepts of creation by natural law, calling it "an intriguing problem in the foundations of modern biology." John F. Cornell, "God's Magnificent Law: The Bad Influence of Theistic Metaphysics on Darwin's Estimation of Natural Selection," *Journal of the History of Biology* 20 (1987): 381–412, at p. 384. See also Dov Ospovat, *The Development of Darwin's Theory: Natural History, Natural Theology and Natural Selection, 1838–1859* (Cambridge: Cambridge University Press, 1981).

biology from the time of its modern foundation. Evolutionary biologists have continued to use it as Darwin did – first, to defeat the argument from design, and second, to extend evolutionary biology's purview, to explain the origins and meaning of human behavior and experience, including those traditionally seen as being outside the domain of scientific or rational explanation, particularly religion.

The philosopher Paul A. Nelson has recently analyzed the first of these problems, the retention of design arguments in modern biology, perspicaciously criticizing on philosophical grounds two of the standard Darwinian strategies used to disprove the argument from design: the argument from imperfection or suboptimal design (appendixes); and the argument for common descent on the basis of homology (wings and flippers). He points out the inherent conflict involved in using a theological strategy to argue for the primacy of methodological naturalism, pointing out that both of these arguments against the argument from design depend upon a number of a priori, underived theological assumptions about the nature of God; for example, the argument from suboptimal design (the panda's thumb, to take Stephen Jay Gould's famous example) rests upon the undemonstrable assumption of an independently derivable optimal design. Nelson concludes that these arguments are a weakness in current evolutionary biology, and recommends that "evolutionary theorists should reconsider both the arguments and the influence of Darwinian theological metaphysics on their understanding of evolution."[17]

Nelson does not consider, however, why it is that this Darwinian natural atheology still has such hold in evolutionary biology, taking it rather as a given, an evolutionary rudiment, as it were. But there are reasons, particularly in Anglo-American evolutionary biology, why this strategy should still hold such appeal for those scientists who are interested in advancing a scientistic, militantly rational metaphysics and who live in a predominantly religious society basing its faith on the principle of revelation. The argument against design can be designed for one of two ends.

In the milder form – which was (probably) Darwin's or, to take a modern example, Stephen Jay Gould's – it serves to demonstrate the nonnecessity of God as an explanation for living design. The question of who or what breathed life into the primordial form remains, and remains open to supernatural explanation, but no such explanations are needed subsequently. Some creator

---

[17] Paul A. Nelson, "The Role of Theology in Current Evolutionary Reasoning," *Biology and Philosophy* 11 (1996): 493–517, at p. 493.

may have had the design of creating a living universe, but that's the most he, she, or it had to do with it. The second, strong form of the argument against design, currently typified perhaps by Richard Dawkins, rests on far shakier metaphysical ground (as Nelson handily demonstrates), being designed to demonstrate the nonexistence of God altogether. In this form, the universe presents a series of test cases for God, and He fails every one. Design is suboptimal. Benevolence is absent. There is no level of analysis on which selfishness does not rule the interactions of living things. Morality is nothing but a human construct; the best ethics we have are made in spite of our animal nature, not because of or through it.

Both of these forms of the argument can naturally be extended to the roots of human behavior. An argument that demonstrates the nonexistence, or at least the uninvolvement, of God in the universe, which is based on an axiomatic methodological rationalism, is almost bound to take up the origins and nature of religion as a subject. This question is, in principle, detachable from the question of the existence of God or the utility of God as an explanation for natural order or as a first cause: God could have created the universe whether anyone believed it or not, and belief can, of course, exist independent of the existence of God. But in practice, the two are generally conflated, and the crypto-theological arguments against design are extended to the validity of religion both as an ontologically independent entity and as a valid underpinning for human behavior. These arguments, too, have two versions, but the end result is much the same, and so is the aim: to reduce the power that religious thinking has over people individually and over society as a whole, if not to do away with it altogether.

In the first version, religion is nothing but a chimera of human culture, a set of superstitions that propagate themselves through purely Lamarckian cultural inheritance: "The meme for blind faith secures its own perpetuation by the simple unconscious expedient of discouraging rational inquiry."[18] Religion can then be cured through the rigorous application of axiomatic rationalism, and, presumably, if enough people could be infected with a Dawkinsian meme for axiomatic rationalism, religion and its (almost wholly malign, because irrational) power over us and our society would go out bang like a candle. The biologist G. C. Williams, whose 1966 *Adaptation and Natural Selection* crucially helped to refine ideas about the meaning and application of the term "adaptation," has also put himself firmly into this camp, citing particularly – in the recent *Plan and Purpose in Nature* – the work of the primatologist

---

[18] Richard Dawkins, *The Selfish Gene* (Oxford: Oxford University Press, 1976), p. 198.

Sarah Blaffer Hrdy on animal and human infanticide to demonstrate nature's essential amorality.[19]

The second version sees religion as a part of human biological nature, and in this case the argument to do away with it entailed by the argument against design must be rather more sophisticated. Darwin, in *The Descent of Man*, linked the developed human "feeling of religious devotion . . . consisting of love, complete submission to an exalted and mysterious superior, a strong sense of dependence, fear, reverence, gratitude, [and] hope for the future" to the manifestations of canine and primate attachment to keepers and to the application of naturally acquired human powers of reasoning and imagination to the mysteries of existence and causation. These, he said, could account for the "belief in unseen or spiritual agencies," which he took to be a human universal, "wholly distinct from that higher [question]" of whether there exists or existed a singular Creator, which Darwin passed off (a trifle disingenuously, particularly in light of his demolishment of the argument from design, the clearest intellectual argument for the existence of God) by saying, "this has been answered in the affirmative by the highest intellects that have ever lived." In any case, "there is no evidence that man was aboriginally endowed with the ennobling belief in the existence of an Omnipotent God";[20] whatever origins religion had thus must have been natural and biological, irrespective of truths they may subsequently have discovered.

E. O. Wilson has been one of the most earnest modern proponents of such a view of religion; in order to extend Darwin's hegemony over methodologies for explaining human experience – an explicit aim of *Sociobiology* (1975) – he modeled religion as simultaneously both epiphenomenal superstition and Darwinian adaptation. Armed both with a selfish-gene perspective and the selfishness-disguised theories of kin selection and reciprocal altruism, Wilson asserted in *Sociobiology* that "when altruism is conceived as the mechanism by which DNA multiplies itself through a network of relatives, spirituality becomes just one more Darwinian enabling device."[21] Its Darwinian effects need not even disguise themselves as altruism; in a taxonomic chapter detailing various types of aggressive interactions, he included "moralistic aggression," adding, "Human moralistic aggression is manifested in countless forms of

---

[19] George C. Williams, *Plan and Purpose in Nature* (London: Weidenfeld and Nicolson, 1996).

[20] Charles Darwin, *The Descent of Man, and Selection in Relation to Sex* (London: John Murray, 1871; reprinted Princeton, NJ: Princeton University Press, 1981), pp. 65–9.

[21] Edward O. Wilson, *Sociobiology: The New Synthesis* (Cambridge, MA: Harvard University Press, 1975), p. 120.

religious and ideological evangelism, enforced conformity to group standards, and codes of punishment for transgressors."[22]

Neither of these forms need necessarily be a part of human biological nature. But Wilson's vision is bleak; where Dawkins saw hope in our ability to transcend nature by means of intellect and write ourselves worthwhile moral codes that might overcome our genetic ones, Wilson is evidently far more pessimistic about the possibilities. He accordingly constructed an evolutionary historiographic scenario in which superstition came to be written into the genes – and, by implication, what has been written in is not so easily written out. He began with the commonplace hypothesis, found in Darwin's *Descent of Man* as well, that religious cohesion offers one tribe or group a competitive advantage vis-à-vis others. (In doing so, however, he departed from the neo-Darwinian orthodoxy of the 1970s that group selection – which can depend upon quite genuine altruism – is an illusory product of competition for individual fitness.)

Wilson saw the origins of religion in what he calls "quite logical" notions of sympathetic magic in proto- or very early humans, who applied their reasoning faculties to the mysteries of causality around them – again paralleling Darwin, whose dog, seeing a parasol blown about by the wind, barked on the rational belief that an invisible intruder must be moving it. So far, a belief in the supernatural is entirely the product of human reason. But Wilson then moved insidiously to inscribe this behavior deeper than the mind, calling it a "reasonable hypothesis" – a plausible scenario – "that magic and totemism constituted direct adaptations to the environment" – eliding the question of whether this adaptation is merely cultural or in fact biological.[23]

Monotheism Wilson accounted for by means of cultural evolution in pastoral societies. Selective mechanisms, ambiguous as to culture or nature, again come into play, as religions "evolve so as to further the welfare of their practitioners. Because this demographic benefit applies to the group as a whole, it can be gained in part by altruism and exploitation, with certain segments profiting at the expense of others." Wilson then slides to the biological level: "Alternatively, it can arise as the sum of generally increased individual fitnesses." The question of the otherwise unaccountable adherence to religions that are in much of their substance "demonstrably false" he finally referred to the "essentially biological question of the evolution of indoctrinability,"

---

[22] Wilson, *Sociobiology*, p. 243.   [23] Ibid., pp. 559–62.

which provides a selective advantage as the religious reap the benefits of a mindlessly conformist but internally peaceable society – either by individual or by group selection, or by both. In short, something that began with the rational application of human faculties to the difficulties of primitive life ends by being destructively impressed upon the entire species, willy-nilly, by a form of the inheritance of acquired characteristics.[24]

This rather Ptolemaic mechanism, saving the appearances while making cumbersome use of a mechanism skirting uncomfortably close to a Darwinian heresy, demonstrates the importance to Wilson of bringing religion under the jurisdiction of evolutionary biology. Dawkins attempts to banish religion by revealing it as an irrational human epiphenomenon; Wilson reduces it by revealing it as an involuntary and equally irrational human adaptation. But why does it matter so much?

All of these attempts to reduce religion, either to a human epiphenomenon or to a quite involuntary phenomenon of our biological nature, have a familiar ring. They are fully a part of the Enlightenment project of establishing the rule of reason, whose three principles Isaiah Berlin summed up in *The Roots of Romanticism* as: "first, that all genuine questions can be answered, that if a question cannot be answered it is not a question; . . . second, . . . that all these answers are knowable, that they can be discovered by means which can be learnt and taught to other persons; . . . [and third,] that all the answers must be compatible with one another."[25] The clear implication and strong hope of all these evolutionary evangelists – Gould, Dawkins, Wilson, Williams, and others who play the "village atheist" – is that if the principles and methods of Reason were only explained clearly enough, religion would lose the tremendous, almost wholly harmful and destructive power that it has over individuals and over society, a hope very much like those expressed by Condorcet and Diderot and the other *Encyclopédistes*, who also hoped to bring all human experience under the rule of one synthetic system of reason, and who also were engaged in a battle against a church whose unreasoning and unreasonable power held sway over a helpless populace. Ruse remarks approvingly in *Mystery of Mysteries* of the biologist Geoffrey Parker – a good scientist, according to him – that "[t]here are positive reasons to think that cultural values do not figure in Parker's work. . . . [H]e argues that his science

---

[24] This is an uncredited and probably independent version of the so-called Baldwin effect; see R. J. Richards, *Darwin and the Emergence of Evolutionary Theories of Mind and Behavior* (Chicago: University of Chicago Press, 1987), pp. 480–95.

[25] Isaiah Berlin, *The Roots of Romanticism*, ed. Henry Hardy (Princeton, NJ: Princeton University Press, 1999), pp. 21–2.

is a way of clearing out the fears and prejudices and hatreds and superstitions brought on by religion."[26]

The irony, and inherent tension, of evolutionary biology is that this search for rational coherence – for "consilience," as Wilson likes to put it – arises as a natural consequence of the nature of evolutionary argument derived from essentially theological argument. This tension has rarely been confronted by evolutionary biologists,[27] and certainly never resolved, to Nelson's frustration. The biases inherent in the argument have shaped the modern science of evolutionary biology – the retention of teleology and adaptationism made virtually inevitable by the substitution of "natural selection" for "the Creator" is only the most obvious and most frequently attacked example.[28] Many of the strongest evangelical atheists of twentieth-century biology – Julian Huxley and E. O. Wilson, in particular – have simultaneously been intransigent believers in the inherent progressivity of evolution; and *progress* in general means one thing: the inevitable appearance of human beings (and, in Wilson's case, ants) at the pinnacle of biological achievement. This transmutated notion of the *scala naturae* has remained so prevalent that when Daniel W. McShea attempted in 1995 to analyze whether any sort of measurable "progress" had in fact occurred in one relatively insignificant branch of living creation, the metazoan animals, the single solid conclusion he could draw was that too little dispassionate consideration had been paid to the concept itself; so far as it existed, the evidence supported "only agnosticism, indeed it supports an emphatic agnosticism."[29] *Agnosticism*, of course, was T. H. Huxley's word for his theological standpoint, coined in the wake of the *Origin*.

The second irony of evolutionary biology is that this transubstantiation of natural theology into the grand narrative of natural selection has itself taken

---

[26] Ruse, *Mystery of Mysteries*, pp. 208–9.

[27] It is especially ironic that one of Gould's *least* percipient essays deals with Adam Smith, Paley, and Darwin. Stephen Jay Gould, "Darwin and Paley Meet the Invisible Hand," *Natural History* 11 (1990): 8–16.

[28] See, for example, Stephen Jay Gould and Richard C. Lewontin, "The Spandrels of San Marco and the Panglossian Paradigm: A Critique of the Adaptationist Program," *Proceedings of the Royal Society of London*, series B – Biological Sciences, 205 (1979): 581–98. G. C. Williams speculated in 1966, "Perhaps biology would have been able to mature more rapidly in a culture not dominated by Judeo-Christian theology and the Romantic tradition. It might have been well served by the First Holy Truth from the Sermon at Benares: 'Birth is painful, old age is painful, sickness is painful, death is painful . . .'" George C. Williams, *Adaptation and Natural Selection: A Critique of Some Current Evolutionary Thought* (Princeton, NJ: Princeton University Press, 1966), p. 255.

[29] Daniel McShea, "Metazoan Complexity and Evolution: Is There a Trend?" *Evolution* 50 (1996): 477–92, at p. 489.

the place of religion for its evangelists. Richard Dawkins, for example, says (at the beginning of *The Blind Watchmaker*) that given the marvellous order of the natural world, he could not imagine being an atheist before 1859, the date of the *Origin of Species* – the implication, of course, being that he could not imagine being anything else afterward, once Darwin had made it possible to be an "intellectually fulfilled atheist."[30] Dawkins and Wilson, among others, have made the jump from the scientist's perhaps requisite epistemic scientism (to use the terminology of the theologian Mikael Stenmark; *epistemic scientism* is "the view that the only reality that we can know anything about is the one science has access to")[31] to what Stenmark calls *redemptive scientism*, in which science takes on the metaphysical, emotional, and cultural functions of religion and can create a fully satisfying world view. The circle is then complete, from the dismantling of a religion based on the rational faculties – natural theology – to its replacement by a science that reaches for faith.

There is also the question of how representative the metaphysics of these vocal evolutionary evangelists are of evolutionary biologists in general. This is a difficult question to answer, of course, in part because of the social and political context of Anglo-American evolutionary biology, which prevents concession or displays of weakness to a fundamentalist enemy; the answer would require extensive knowledge of opinions carefully kept private.

The biologist Theodosius Dobzhansky discussed the relationship between his Christian faith and his science in an exchange of letters with the philosopher John Greene in the early 1960s. He wrote that he believed Julian Huxley's antireligious stance to be "very much a majority opinion" among natural scientists but that he himself was unable to subscribe to it; at the same time, he found aspects of his two faiths irreconcilable. Evolution was for him "a bright light. But it does not follow that evolution is a source of natural theology and a 'proof' of the existence of God. . . . I am groping for a tolerable self-consistent Weltanschauung but do not claim having found one."[32] It may well be that many, if not most, evolutionary biologists are, like Dobzhansky, quiet theists of one kind or another, and if so, this would provide a larger context for understanding visions and revisions of Darwin's natural atheology in twentieth-century biology.

---

[30] Richard Dawkins, *The Blind Watchmaker* (New York: Norton, 1986), p. 6.
[31] Mikael Stenmark, "What is Scientism?," *Religious Studies* 33 (1997): 15–32, at p. 19.
[32] John C. Greene and Michael Ruse [and Theodosius Dobzhansky], "On the Nature of the Evolutionary Process: The Correspondence between Theodosius Dobzhansky and John C. Greene," *Biology and Philosophy* 11 (1006): 445–91, at p. 463.

Finally, all of these narratives meant to explain the shape of life on Earth are meant also to explain something much more specific. If they did not explain us – why we are here and what we mean – they would not have the power they do. The issue of human exceptionalism – do humans share exactly the same status, whatever that means, as all other living things, or are we in some real way special? – is central, though tacit, for all these thinkers, from Paley and Darwin to now, who have tried to synthesize a broad picture of the history of life. In general, of course, the answer has been yes: we, or something very like us, have been in some way a foreordained result of evolution. Of the many criteria inherited by evolutionary biology from theology, this is surely the most important. The a priori belief in human exceptionalism provides one of the strongest, if unstated, criteria for judging the plausibility of evolutionary narratives.[33] If we are no longer made in God's image, then in whose can we cast ourselves? In Darwin's?

[33] See also McShea, "Metazoan Complexity," particularly p. 488.

CHAPTER FIVE

Ironic Heresy

How Young-Earth Creationists Came to Embrace Rapid Microevolution by Means of Natural Selection

Ronald L. Numbers

Some years after writing his famous essay *On the Origin of Species* (1859), Charles Darwin noted that his primary goals had been to overthrow "the dogma of separate creations" and to establish natural selection as the primary, through far from exclusive, mechanism of change. Regarding the relative importance of these twin goals, he left no doubt. "Personally, of course, I care much about Natural Selection," he confided to an American correspondent; "but that seems to me utterly unimportant, compared to the question of Creation or Modification." Well into the twentieth century naturalists continued to debate the merits of natural selection, but since the early 1870s they have been describing the theory of common descent as an "ascertained fact." The ultimate Darwinian heresy was thus the denial of common descent.[1]

Despite the frequent claims of anti-evolutionists to the contrary, during the first quarter of the twentieth century about the only biologist of repute who

___

[1] Charles Darwin, *The Descent of Man, and Selection in Relation to Sex*, 2 vols. (London: John Murray, 1871), vol. 1, pp. 152–3; Charles Darwin to Asa Gray, May 11, 1863, quoted in Francis Darwin, ed., *The Life and Letters of Charles Darwin*, 2 vols. (New York: Appleton, 1896), vol. 2, pp. 163–4; E. D. Cope, *The Origin of the Fittest* (New York: Appleton, 1887), p. 2, from "Evolution and Its Consequences," first published in *Penn Monthly Magazine* in 1872. On the responses of American naturalists to organic evolution, see Ronald L. Numbers, *Darwinism Comes to America* (Cambridge, MA: Harvard University Press, 1998), pp. 24–48.

Much of the material in this essay was previously used in my book *The Creationists* (New York: Knopf, 1992). I would like to thank my research assistant, Spencer Fluhman, for his help in the preparation of this paper, and the participants in the Darwinian Heresies conference at the Max Planck Institute in Berlin, December 15–16, 2000, for their suggestions.

rejected organic evolution was Albert Fleischmann (1862–1942), a respectable but relatively obscure German zoologist who taught for decades at the University of Erlangen in Bavaria. In 1901 he published a scientific critique of organic evolution, *Die Descendenztheorie*, in which he dismissed not only natural selection but also the notion of common descent. This gave him a unique reputation among biologists. As Vernon L. Kellogg noted in 1907, Fleischmann seemed to be "the only biologist of recognized position . . . who publicly declared a disbelief in the theory of descent." The German anti-evolutionist apparently remained of the same mind for the rest of his life. In 1933, the year of his retirement from Erlangen, he presented a paper to the Victoria Institute in London in which he dismissed the notion of a "genealogical tree" as a "fascinating dream." "No one can demonstrate that the limits of a species have ever been passed," he asserted. "These are the Rubicons which evolutionists cannot cross." In his declining years, Fleischmann informed English acquaintances that he was writing a book "that will wipe evolution off the slate," but the work never appeared.[2]

During the 1920s and 1930s, creationists sometimes tried to bolster their position by allying themselves with noncreationist critics of Darwinism, only to discover that the differences between them were greater than the similarities. British anti-evolutionists for a time became enamored of the French zoologist Louis Vialleton (1859–1929), a professor of comparative anatomy at the University of Montpellier, who rejected the notion of mechanistic, continuous development in the organic world. But despite his skepticism, he remained an *évolutionniste*.[3] Similarly, American anti-evolutionists cheered when the distinguished Smithsonian zoologist Austin H. Clark (1880–1954), struck by the absence of intermediate forms between the major groups of animals, challenged the conventional view of evolution that represented species as branches of a single tree. But both in correspondence with creationists and

[2] Albert Fleischmann, *Die Descendenztheorie* (Leipzig: Verlag von Arthur Georgi, 1901); Vernon L. Kellogg, *Darwinism To-day* (New York, Henry Holt, 1907), p. 8; Albert Fleischmann, "The Doctrine of Organic Evolution in the Light of Modern Research," *Journal of the Transactions of the Victoria Institute* 65 (1933): 194–214, at pp. 196, 205–6; Douglas Dewar to [name deleted], November 2, 1931, from a copy in the George McCready Price Papers, Adventist Heritage Center, Andrews University. For bibliographical information, see George Uschmann, "Albert Fleischmann," *Neue Deutsche Biographia*, vol. 5 (Berlin: Duncker and Humblot, 1960), pp. 234–5.

[3] Harry W. Paul, *The Edge of Contingency: French Catholic Reaction to Scientific Change from Darwin to Duhem* (Gainesville: University Presses of Florida, 1979), pp. 99–101; Douglas Dewar, "The Limitations of Organic Evolution," *Journal of the Transactions of the Victoria Institute* 64 (1932): 120–43; Douglas Dewar, *More Difficulties of the Evolution Theory: And a Reply to "Evolution and Its Modern Critics"* (London: Thynne, 1938), pp. 101–3, 137–8.

in his controversial book *The New Evolution: Zoogenesis* (1930), Clark point-edly refused to support a supernatural view of origins. As one confused and disappointed creationist explained, "Dr. Clark . . . denies the most pointed evidence of evolution, but he sticks to evolution just the same."[4]

Unfortunately for the American crusaders against evolution in the 1920s, not one of them possessed an advanced degree in biology. Among the four leading scientific spokesmen, the Canadian surgeon Arthur I. Brown (1875–1947), often touted as the greatest "scientist" in the anti-evolution camp, held only medical degrees. Harry Rimmer (1890–1952), a Presbyterian minister who billed himself as a "research scientist," had picked up his scientific vocab-ulary while attending a small homeopathic medical school. George McCready Price (1870–1963), recognized by *Science* as "the principal scientific authority of the Fundamentalists," had acquired his scientific knowledge by extensive reading. S. James Bole (1875–1956), a professor of biology at fundamentalist Wheaton College in Illinois, had devoted years to the study of pomology (i.e., fruit culture) as a graduate student in the agricultural school of the University of Illinois, but his only graduate degree at the time was an A.M. in education, from Illinois, for which he had submitted a thesis on elementary-school pen-manship. It was not until 1934 that he earned a Ph.D. in horticulture from Iowa State College in Ames.[5]

Most of the early anti-evolutionists instinctively defended the fixity of species. Like the great eighteenth-century Swedish taxonomist Carolus Linnaeus (1707–1778), they assumed that "just so many species are to be reckoned as there were forms created in the beginning." Rimmer, in a typical statement, explicitly equated Genesis "kinds" with "species" and insisted that all the varieties of a species had descended from "a common ancestral pair."[6]

---

[4] Austin H. Clark, *The New Evolution: Zoogenesis* (Baltimore: Williams and Wilkins, 1930); Ben F. Allen to G. M. Price, June 12, 1929, Price Papers. See also Austin H. Clark to G. M. Price, March 23, 1929, Price Papers; and Theodore Graebner, *God and the Cosmos: A Critical Analysis of Atheism* (Grand Rapids, MI: Eerdmans, 1932), pp. 287–9.

[5] For biographical information about Brown, Rimmer, Price, and Bole, see Ronald L. Numbers, *The Creationists* (New York: Knopf, 1992), pp. 54–101.

[6] Arthur I. Brown, *Evolution and the Bible* (Vancouver: Arcade Printers, [1922]), p. 17; Harry Rimmer, *Modern Science, Noah's Ark, and the Deluge* (Los Angeles: Research Science Bureau, 1925), p. 10; Harry Rimmer, *The Facts of Biology and the Theories of Evolution* (Los Angeles: Research Science Bureau, 1929), p. 10. The fixity of species was also commonly defended by the leading clerical anti-evolutionists. See, e.g., Alexander Patterson, *The Other Side of Evolution: Its Effects and Fallacy* (Chicago: Bible Institute Colportage Assn., 1903), p. 26; L. T. Townsend, *Collapse of Evolution* (Boston: National Magazine Co., 1905), p. 21; John Roach Straton and Charles Francis Potter, *Evolution versus Creation: Second in the Series of Fundamentalist–Modernist*

Such expressions left an indelible impression and prompted Theodosius Dobzhansky, for example, to write in 1973 that anti-evolutionists "fancy that all existing species were generated by supernatural fiat a few thousand years ago, pretty much as we find them today."[7] As we shall see, he should have known better. By the 1970s, most of the leading special creationists had long since abandoned belief in the fixity of species and had embraced extensive – and extremely rapid – organic evolution within the originally created "kinds," mentioned in the first chapter of Genesis.

Although not a biologist, Price penned the first extensive treatments of speciation to come from a twentieth-century American creationist. In *Q.E.D.; or, New Light on the Doctrine of Creation* (1917) he devoted an entire chapter to answering the question "What Is a 'Species'?" Best known for fathering the theory of flood geology (later renamed creation science), Price restricted the history of life on earth to about 6,000 years and assigned most of the fossil-bearing rocks to the year-long cataclysm associated with the biblical Noah. As a devout Seventh-day Adventist, he believed in the divine inspiration of the Adventist prophetess Ellen G. White (1827–1915), whose visionary experiences allowed her to write authoritatively on topics ranging from biology and geology to eschatology and soteriology. At times her influence on his writings (and on the views of other Adventist creationists) proved decisive.[8]

Price had little quarrel with evolutionists regarding the origin of *species*, as biologists commonly used the term. Although he preferred to equate "species" with the originally created Genesis "kinds," and in later life regretted that he had earlier conceded so much to evolutionists, he at times freely admitted that new species, narrowly defined, had evolved from "the great stocks, or families" created by God – and at a rate far faster than most evolutionists demanded. The more variation he allowed, the easier it was for him to sidestep "a real difficulty," namely, explaining "how the great diversity of our modern world may have come about after the world disaster of the Deluge, from a comparatively few kinds which were salvaged from that great cataclysm." In

Debates (New York: George H. Doran, 1924), p. 100; W. B. Riley, *Darwinism; or, Is Man a Developed Monkey?* (Minneapolis: the author, [1929]), p. 8.

[7] Theodosius Dobzhansky, "Nothing in Biology Makes Sense Except in the Light of Evolution," *American Biology Teacher* 35 (1973): 125–9, at p. 127.

[8] George McCready Price, *Q.E.D.; or, New Light on the Doctrine of Creation* (New York: Fleming H. Revell, 1917), pp. 68–77. On Price, see Numbers, *The Creationists*, pp. 72–101. On White, see Ronald L. Numbers, *Prophetess of Health: A Study of Ellen G. White* (New York: Harper and Row, 1976).

1925, he justified his views in a revealing statement to his fellow Adventist science teachers:

Personally, I believe that these great family types are the ones that were originally created, and that a false issue has been raised over the "origin of species." . . . I think it is quite reasonable to suppose that all our cats are of one stock, that all our cattle are of a common origin, and that all the dogs and wolves may be of a common descent. To suppose this is only to suppose something which helps us to see how the great diversity around us may have come about from a comparatively few original stocks which survived the great world disaster which the Bible and a rational geology alike declare has actually taken place.

The point at issue between creationists such as Price and evolutionists was not variation but its extent and direction: "whether the general run of these changes have not all been in the direction of degeneration, not development."[9]

As so often was the case, Price found guidance in dealing with the species question in the writings of White, who in 1864 had written:

Every species of animal which God had created were preserved in the ark [of Noah]. The confused species which God did not create which were the result of amalgamation, were destroyed by the flood. Since the flood there has been amalgamation of man and beast, as may be seen in the almost endless varieties of species of animals, and in certain races of men.[10]

Although understandably reticent to discuss White's divinely inspired views in his published works, especially those intended for a non-Adventist audience, he privately subscribed to the idea that Satan himself, "the great primal hybridizer," had been "the real instigator of all the mixing and crossing of the races of mankind, and also the mixer of thousands of kinds of plants and animals which God designed should remain separate." Thus Price substituted demonic manipulation for natural selection in order to explain the origin of species.[11]

Publicly, Price tended to attribute the appearance of new species to environmental influences, the "*vital elasticity*" of organisms, and *divine* intervention. In a chapter titled "Species and Their Origin" in *The Phantom of Organic Evolution*, he outlined a theory of common descent with modifications

[9] Price, *Q.E.D.*, pp. 68–77; Price, *Outlines of Modern Christianity and Modern Science* (Oakland, CA: Pacific Press, 1902), p. 199; Price, "Dear Fellow Science Teachers," *Watchman Magazine* 34 (January 1925): 18. On the identification of families with the originally created kinds, see also Price, *The Phantom of Organic Evolution* (New York: Fleming H. Revell, 1924), pp. 212–17.
[10] Ellen G. White, *Spiritual Gifts: Important Facts of Faith, in Connection with the History of Holy Men of Old* (Battle Creek, MI: Seventh-day Adventist Publishing Assn., 1864), p. 75.
[11] G. M. Price to Martin Gardner, May 13, 1952, Gardner Papers, courtesy of Martin Gardner.

characterized by a process of "degeneration downward" rather than "development upward" and by quick rather than slow change. The most rapid change
had probably occurred in the immediate post-deluge period, when both humans and animals had encountered a novel environment. In a textbook for
high school students he explained the process:

Very radical changes in the environment of plants or of animals tend to make the
species vary; and if they survive in such new environment, their size or color or habits,
or other physical "characters," will be different. This is the reason why the living forms
are so different, in some instances, from their ancestral forms found as fossils in the
rocks; for there has been a very radical change in their environment in passing from
the antediluvian world to the modern one. Some species have changed so considerably
that scientists do not recognize them as the same, but give them different specific or
even different generic names, calling the older form "extinct," and the modern form
a "new" species.

According to Price, new postdiluvian species arose not so much by means
of Darwinian natural selection or by a Lamarckian inheritance of acquired
characteristics but because "the great superintending Power which is over nature, adapted these men and these animals and plants to their strange world."
In accounting for the geographical distribution of plants and animals, which
he admitted was a "very difficult" problem for creationists because nearly
identical environments did not always possess the same flora and fauna, he
similarly appealed to divine intervention, speculating that after the deluge
animals spread out from Mount Ararat "under the direct guidance of the
Creator." Rapid speciation ensued.[12]

One of the first twentieth-century anti-evolutionists to obtain a graduate degree in biology was Harold W. Clark (1891–1986), a former student
of Price's. He not only wrote the earliest book, *Back to Creationism* (1929),
explicitly promoting "creationism" (rather than anti-evolutionism), but also
authored a creationist classic, *The New Diluvialism* (1946), which attributed
the geological strata to the rapid burial of antediluvian ecological zones during
Noah's flood.[13] Clark spent the 1932–33 academic year studying for a master's

---

[12] Price, *Q.E.D.*, p. 90 (vital elasticity); Price, *Phantom of Organic Evolution*, pp. 91–112; Price, *A
Textbook of General Science for Secondary Schools* (Mountain View, CA: Pacific Press, [1917]),
pp. 500–10; Price, *The Geological-Ages Hoax: A Plea for Logic in Theoretical Geology* (New
York: Fleming H. Revell, 1931), pp. 105–6.

[13] Harold W. Clark, *Back to Creationism* (Angwin, CA: Pacific Union College Press, 1929); Clark,
The New Diluvialism (Angwin, CA: Science Publications, 1946). On the use of the term
"creationism," see Ronald L. Numbers, "Creating Creationism: Meanings and Uses since the
Age of Agassiz," in David N. Livingstone, D. G. Hart, and Mark A. Noll, eds., *Evangelicals and
Science in Historical Perspective* (New York: Oxford University Press, 1999), pp. 234–43.

degree in biology at the University of California. After completing his studies at Berkeley, he updated and enlarged *Back to Creationism*, trying to make it as scientifically respectable as possible within the constraints of a 6,000-year history. He pointedly challenged the common creationist notion that there had been "no change in species since the beginning." Hybridization, he maintained, had been "a very potent factor in the formation of new species." When Price read the manuscript, he enthusiastically endorsed Clark's acceptance of "species-making by means of hybridization and otherwise," noting that these suggestions corroborated White's puzzling statement about amalgamation between humans and animals. In 1940, Clark published the biological part of his manuscript under the title *Genes and Genesis* (1940), in which he defended limited Darwinian natural selection – within genera, families, and even orders – against the "extreme creationism" of anti-evolutionists who still insisted that God had created every species.[14]

Although some creationists sided with Clark – welcoming, as one of them put it, "every bit of *modification during descent* that can reasonably be asserted" in order to avoid making Noah's ark "more crowded than a sardine can" – others resisted the effort to embrace the natural evolution of species. Byron C. Nelson (1893–1972), a conservative Lutheran minister who had written *"After Its Kind": The First and Last Word on Evolution* (1927), worried that such concessions had "given up half of the battle" to the evolutionists. Clark and his supporters were throwing in the towel, said Nelson, because they failed to recognize the immense size of the ark and the small number of "large, bulky species like elephants, rhinocerouses [sic], lions, horses, cattle etc." that had to be accommodated. All of the large animals, Nelson surmised, could be "put on one floor easily," and the smaller species, such as rabbits, squirrels, and birds, did not take much room. But even while defending the fixity of all "natural species," Nelson granted that "new *varieties* may have arisen since the creative days." In the end, his quarrel with Clark amounted to little more than a semantic disagreement. Nelson may have insisted on the fixity of "species," but he allowed the originally created units to encompass entire taxonomic

---

[14] H. W. Clark to officers of the Ministerial Association, March 10, 1937, Price Papers (hybridization); G. M. Price to M. E. Kern and others, ca. 1936, Papers of the Publishing Department of the General Conference of Seventh-day Adventists, Silver Spring, MD; H. W. Clark, *Genes and Genesis* (Mountain View, CA: Pacific Press, 1940), pp. 43, 143; G. M. Price, "The 'Amalgamation' Question Again," unpublished ms., ca. 1941, Ballenger Papers, courtesy of Donald F. Mote (shame). See also "Statement of Prof. H. W. Clark re Amalgamation," undated ms., Publishing Department Papers; author's interview with Harold W. Clark, May 11, 1973; Harold W. Clark, *Genesis and Science* (Nashville, TN: Southern Publishing Assn., 1967), p. 57.

families. He conceded, for instance, that "all the 'cats' throughout the world are one natural species, descended from one common pair."[15]

In the early 1940s, another university-trained Price protégé, Frank Lewis Marsh (1899–1992), joined Clark in advocating post-Edenic speciation. While teaching at an Adventist school in the Chicago area, Marsh took advanced work in biology at the University of Chicago and obtained an M.S. in zoology from Northwestern University in 1935, specializing in animal ecology. Later, after joining the faculty of the Adventists' Union College in Lincoln, he completed a Ph.D. in botany at the University of Nebraska in 1940, where he wrote his dissertation on plant ecology and became the first Adventist (and one of the first creationists) to earn a doctoral degree in biology.[16]

Like Clark, Marsh never deviated from a literal, recent creation and universal flood, but the more he learned, the more he questioned the notion that all species had originated by separate creative acts. Zoologists, he noted, had identified thousands of species of dry-land animals alone, yet Adam had been able to name all of them in less than a day. Thus it seemed unreasonable to equate the Genesis kinds with the multitudinous species of the twentieth century. There was also the troublesome matter of limited space on Noah's ark. Because Ellen White had written under divine inspiration that "[e]very species of animal which God had created were preserved in the ark," it behooved her followers to keep the number of originally created "species" to a minimum. For personal reasons, Marsh trusted evolutionary biologists more than some of his fundamentalist colleagues. As he once explained to Price, his close association with evolutionists over the years had given him "an understanding of their way of thinking" and a confidence in their taxonomic work that Price could never appreciate. "You have never rolled up your sleeves and worked as one of their crowd on various research projects as I have," he reminded the self-taught geologist.[17]

---

[15] D. J. Whitney to B. C. Nelson, June 3 and June 18, 1928, and Nelson to Whitney, April 16, 1929, all in the Byron C. Nelson Papers, courtesy of Paul Nelson; Byron C. Nelson, *"After Its Kind": The First and Last Word on Evolution* (Minneapolis: Augsburg Publishing House, 1927), pp. 18–24. On Nelson, see Paul Nelson's biographical introduction to his edited volume, *The Creationist Writings of Byron C. Nelson*, vol. 5 of Ronald L. Numbers, ed., *Creationism in Twentieth-Century America: A Ten-Volume Anthology of Documents, 1903–1961* (New York: Garland, 1995).

[16] Frank L. Marsh to Theodosius Dobzhansky, February 21, 1945, Frank L. Marsh Papers, Adventist Heritage Center, Andrews University. Biographical data come from the author's interview with Marsh, August 30, 1972; Marsh to Ronald L. Numbers, April 10, 1974; and Marsh, "Life Summary of a Creationist," unpublished ms., December 1, 1968, Marsh Papers.

[17] Frank Lewis Marsh, *Evolution, Creation, and Science* (Washington, DC: Review and Herald Publishing Assn., 1944), pp. 165–6; Marsh to Price, September 5, 1943, Price Papers.

In his first book, *Fundamental Biology* (1941), written from the point of view of a "fundamentalist scientist," Marsh portrayed the living world as the scene of a cosmic struggle "between the Creator and Satan." Confusion about the nature of the originally created "kinds" often arose, he warned, because Satan, "a master geneticist," had "built up within the kinds, different races, strains, and types which look quite unrelated to other members of the kind." Taking his cue from White, whom he regarded "as essential to the biologist today as an aid in understanding the written Word and the book of nature," Marsh speculated that amalgamation or hybridization had been "the principal tool used by Satan in destroying the original perfection and harmony among living things." It explained "the origin of multitudinous species, varieties, and races of plants and animals on the earth today." The black skin of Negroes was only one of many "abnormalities" engineered in this diabolical way.[18]

Although in agreement with Clark on the natural origin of species, he thought his Adventist colleague had gone too far in allowing for the cross-breeding of "kinds," which Marsh chose to call "baramins" (from the Hebrew words *bara*, "created," and *min*, "kind") and which he later identified with the "polytypic species" of Ernst Mayr (b. 1904). In contrast to Clark, who believed that the biblical reference to plants and animals producing "after their kind" was merely a moral principle, Marsh regarded the Genesis statement as a biological law that forever separated the different "kinds." The examples of interbreeding between "kinds" that Clark gave in *Genes and Genesis* struck Marsh as being more appropriate to *Ripley's Believe It or Not* than to a scientific treatise. And he was sure that Clark's proof of human–beast crossings, cited as confirmation of White, "would not last five minutes under a scientific cross examination." Such lapses he attributed to the fact that Clark had studied only at Berkeley, while he himself had been privileged to take advanced work at *three* institutions of higher learning: Chicago, Northwestern, and Nebraska.[19]

Despite receiving Price's imprimatur, Marsh encountered severe criticism from some fellow creationists, especially from members of the Deluge Geology Society, founded in 1938 to promote Price's reading of Earth history. After sending society members the manuscript of an updated and expanded version of his views, published as *Evolution, Creation, and Science* (1944), Marsh found himself having to squelch "the wild rumor which comes to me that some of you

---

[18] Frank Lewis Marsh, *Fundamental Biology* (Lincoln, NE: the author, 1941), pp. iii, 6, 11, 48, 56, 63.

[19] Excerpts from letters written by F. L. Marsh to D. E. Robinson, February 16 and March 16, 1941, Publishing Department Papers. On "baramins," see Marsh, *Fundamental Biology*, p. 100; and Marsh, *Evolution, Creation, and Science*, p. 162. On the identifying of baramins with polytypic species, see Marsh to Dobzhansky, January 12, 1945, Marsh Papers.

men think I am an evolutionist." He begged his "brother fundamentalists" not to equate limited variation with evolution. "It is the constant refusal on the part of many special creationists to recognize the change in nature which actually does occur which has led [H. H.] Newman [one of Marsh's teachers at the University of Chicago] to dub us 'ignorant, dogmatic, or prejudiced' – and I think he is justified in saying just that!" When the book finally appeared, one creationist geneticist accused him of being "much more of an evolutionist than any man I have ever known." In reviewing the work at a meeting of the Deluge Geology Society, the geneticist portrayed Marsh as a man "willing to grant everything that evolutionists ask for and boil it down to just about five thousand years, on a scale that they don't even dream of in hundreds of thousands, or even millions, of years."[20]

Beginning with *Evolution, Creation, and Science*, Marsh sanitized his major published works in order to avoid mentioning "the Spirit of Prophecy" (that is, Ellen G. White or her writings), which he feared would only repel non-Adventist readers. Hoping that a defense of creationism written by a credentialed biologist might cause the scientific world to take note, he asked his publisher to send out complimentary copies to prominent evolutionists, including the Harvard zoologist Ernst Mayr and the Russian-born Columbia geneticist Theodosius Dobzhansky (1900–1975). The former declined to comment, preferring to take the advice of "the rabbit in Walt Disney's film *Bambi*: 'Don't say anything if you don't have anything nice to say.'" Dobzhansky, however, believing that "the majority should at least consider the minority view and subject it to criticism," engaged in an extended correspondence with Marsh (from November 1944 to February 1945) that vividly reveals not only the issues diving creationists and evolutionists but also the extent to which the former were adopting limited evolution.[21]

Shortly before the appearance of Marsh's book, the Russian Orthodox émigré had remarked in his *Genetics and the Origin of Species* (1937) that "among the present generation no informed person entertains any doubt of

[20] G. M. Price to B. F. Allen, February 13, 1944 (fool); F. L. Marsh, "Confessions of a Biologist," unpublished ms., August 25, 1943 (wild rumor); Walter Lammerts and others, "Review of 'Creation, Evolution, and Science,'" by Frank L. Marsh [November 18, 1944]; all in the Couperus Papers, courtesy of the late Molleurus Couperus. The description of the meeting appears in *Creation-Deluge Society News Letter*, December 16, 1944. Marsh identifies Newman as his teacher in a letter to Dobzhansky, December 13, 1944, Marsh Papers.

[21] Marsh to Price, October 16, 1947 (sanitized); Mayr to Marsh, March 13, 1945; both in the Price Papers. Dobzhansky to Marsh, November 29, 1944, Marsh Papers. Marsh's later books included *Studies in Creationism* (Washington, DC: Review and Herald Publishing Assn., 1950); *Life, Man, and Time* (Mountain View, CA: Pacific Press, 1957); *Evolution or Special Creation?* (Washington, DC: Review and Herald Publishing Assn., 1963); and *Variation and Fixity in Nature* (Mountain View, CA: Pacific Press, 1976).

the validity of the evolution theory in the sense that evolution has occurred."
But reading Marsh's book convinced him otherwise. In reviewing it for the
American Naturalist, Dobzhansky announced that Marsh had written what he
had previously thought to be impossible: a sensibly argued defense of special
creation. Dobzhansky expressed particular surprise at discovering how much
evolution (within "kinds") a creationist such as Marsh apparently was willing
to grant: "He outbids evolutionists on the score of the speed of the changes,
for he assumed that all dogs, foxes, and hyenas are members of a single 'kind,'
and, therefore, must have descended from a common ancestor in any event less
than 6000 years ago." But in rejecting macroevolution, Marsh's book taught
the valuable lesson that "no evidence is powerful enough to force acceptance
of a conclusion that is emotionally distasteful."[22]

Although Dobzhansky found Marsh's ideas scientifically invalid and religiously subversive, because they implied a deceptive Creator, he grudgingly
respected the church-college biologist for being "the only living scientific anti-
evolutionist." The obscure creationist, for his part, could scarcely conceal his
delight at having the unexpected chance to argue the case for creationism before one of the leading evolutionists in the world. In response to Dobzhansky's
comment about outbidding the evolutionist, Marsh amended and clarified his
position. Breaking with "the Linnaean position," which equated species with
the original units of creation, he explained that God in originally stocking the
world with plants and animals had created not only "kinds" but also "races"
or "varieties" within those kinds. Thus, for example, rather than insisting that
all dogs, foxes, and hyenas had come from a single ancestor, he suggested
that perhaps they had descended from three originally created varieties of the
"dog kind," all capable of interbreeding and each represented on Noah's ark:
"a fox-like 'variety,' a dog-like 'variety,' and a hyena-like 'variety.'" Although
he had said virtually nothing about the mechanism of change in his book, he
now admitted that the concepts of natural selection and survival of the fittest
were "really extremely important" in explaining the present distribution of
species.[23]

The central issue separating the two biologists hinged on the nature of
scientific proof. Marsh, who assimilated all the evidence of microevolution

[22] Theodosius Dobzhansky, *Genetics and the Origin of Species* (New York: Columbia University Press, 1937), p. 8; Dobzhansky, review of *Evolution, Creation, and Science*, by F. L. Marsh, *American Naturalist* 79 (1945): 73–5. See also Dobzhansky to Marsh, November 15, 1944, Marsh Papers.

[23] Marsh to Dobzhansky, November 19, 1944 (delight); Marsh to Dobzhansky, December 4, 1944 (varieties and natural selection); Dobzhansky to Marsh, December 7, 1944 (only anti-evolutionist); all in the Marsh Papers.

into his creationist paradigm of changes within "kinds," demanded nothing less than laboratory-based demonstrations of macroevolution. But, as Dobzhansky pointed out, the evidence for such large-scale evolution rested on inference, not direct observation. Because macroevolution took place in geological time, he patiently explained, it could "be proven or disproven only by inference from the available evidence." Marsh, predictably, found this argument unconvincing. "Alas! Inferential evidence again!" he exclaimed. "Is there no *real* proof for this theory of evolution which we may grasp in our hands?" Eventually, explanation gave way to frustration, with Dobzhansky finally brushing Marsh's concern aside with the quip, "If you demand that biologists would demonstrate the origin of a horse from a mouse in the laboratory then you just can not be convinced."[24]

As Marsh readily admitted, he rejected the inferential evidence for macroevolution primarily because of his prior commitment to "the literality of biblical statements." In justifying such an allegiance, he maintained that "in not one single instance" had the Bible been proven false. "That very real fact should mean something to us as scientists," he argued. "In the light of that fact, the Genesis statement regarding the origin of living things must likewise be tested if we are to make wise use of the sources of truth at our disposal." Dobzhansky found this line of reasoning unpersuasive, but he credited Marsh for his candor in stating that "the account given by the Bible is settled for you before you begin to consider the biological evidence." To Marsh's annoyance, Dobzhansky refused to concede that evolutionists were equally under the influence of preconceived notions – that, in Marsh's words, "they are sold heart and soul to their theory and are even ready, in order to accomplish it, to exchange kinship with the great God for bloodrelationship with apes." Dobzhansky not only denied harboring any preconceived sentiments against creationism but also asserted that he "would just as well see creation as evolution theory prove right" – if only the facts would allow it.[25]

After more than two months of almost weekly exchanges, Marsh trusted Dobzhansky sufficiently to expose some of his innermost fears and feelings. In what he suspected would be his final letter, he assured the geneticist that he was not "a chronic grouch who goes about looking for something to argue about." Nor was he "looking for ease, comfort, and a good name." Though he

---

[24] Dobzhansky to Marsh, December 22, 1944 (geological time); Marsh to Dobzhansky, January 12, 1945 (real proof); Dobzhansky to Marsh, February 5, 1945 (mouse); all in the Marsh Papers.

[25] Dobzhansky to Marsh, December 22, 1944, and February 5, 1945; Marsh to Dobzhansky, January 12, 1945; all in the Marsh Papers.

disliked being at odds with his scientific brethren, he was "willing to take it on the chin," if principle required it, if only he could get mainstream scientists to accept special creation as a legitimate alternative to evolution. In closing, he expressed the hope that Dobzhanksy would find "some diversion in these letters, some pleasant mental gymnastics, and possibly experience a broader acquaintance with unusual ideas so that the benefits will not all be going one way." Six years later, in the third edition of his *Genetics and the Origin of Species*, Dobzhansky cited Marsh as the exception to the rule that "an informed and reasonable person can hardly doubt the validity of the evolution theory, in the sense that evolution has occurred." The creationist biologist proved "only that some people have emotional biases and preconception[s] strong enough to make them reject even completely established scientific findings." It wasn't much of an acknowledgement, but Marsh appreciated the recognition.[26]

Early in 1945, the *Bulletin of Creation, the Deluge and Related Science* published a glowing review by Marsh of the second edition of Dobzhansky's *Genetics and the Origin of Species*. Marsh praised the author's "sincerity, analytical power, and nearly faultless logic." Marsh's warm feelings were not reciprocated. Although Dobzhansky himself became increasingly religious in later life – possibly as a result of his growing friendship with the deeply religious Australian zoologist Charles Birch and his fondness for the Jesuit priest Pierre Teilhard de Chardin's *The Phenomenon of Man* (1963) – his exchange with Marsh had left him with a jaundiced view of creationists as inflexible ideologues. Over twenty years after his exchange with Marsh, when sharing his own views on philosophy and religion in *The Biology of Ultimate Concern* (1967), Dobzhansky noted "the really extraordinary phenomenon" of

the continued existence of a small minority of scientifically educated fundamentalists who know that their beliefs are in utter, flagrant, glaring contradiction with firmly established scientific findings. . . . I had a futile and exasperating correspondence with one antievolutionist "creationist," who could not be accused of unfamiliarity with the relevant evidence. Discussions and debates with such persons are a waste of time; I suspect that they are unhappy people, envious of those who are helped to hold similar views by plain ignorance.[27]

[26] Marsh to Dobzhansky, February 21, 1945, Marsh Papers; Theodosius Dobzhansky, *Genetics and the Origin of Species*, 3rd ed. (New York: Columbia University Press, 1951), p. 11. Marsh quotes Dobzhansky's mention of him in *Evolution or Special Creation?*, p. 46.
[27] Frank Lewis Marsh, "The Present Status of Genetics and the Origin of Species," *Bulletin of Creation, the Deluge and Related Science* 5 (1945): 1–9, at p. 1; Theodosius Dobzhansky, *The Biology of Ultimate Concern* (New York: New American Library, 1967), pp. 95–6; Charles E. Taylor, "Dobzhansky, Artificial Life, and the 'Larger Questions' of Evolution," in Mark B. Adams, ed., *The Evolution of Theodosius Dobzhansky: Essays on His Life and Thought in*

Creationists such as Clark and Marsh may have encouraged interested fundamentalists to abandon belief in the fixity of species, but they remained
resolutely opposed to natural evolution that transcended the originally created "kinds," however defined. However, halting the evolutionary process at
kinds was not always easy, as members the evangelical American Scientific
Affiliation (ASA) discovered in the 1950s and 1960s. Established as a creationist organization in 1941, the affiliation found itself moving rapidly from
strict creationism to progressive creationism and on to theistic evolutionism. Leading the way toward an accommodation with modern biology was
Russell L. Mixter (b. 1906), a Wheaton College professor of biology who had
earned a Ph.D. in anatomy from the University of Illinois School of Medicine.
Mixter began undermining fundamentalist beliefs in a recent creation and
the fixity of species in papers that he presented at ASA conventions in the late
1940s, subsequently published collectively as *Creation and Evolution* (1950).
In this monograph, which enjoyed wide use in evangelical colleges and sem-
inaries, he dismissed Noah's flood as an event of no biological consequence
and advocated the acceptance of evolution "within the order." Progressive
creationists, he concluded, could "believe in the origin of species at different
times, separated by millions of years, and in places continents apart."[28] On
one occasion Mixter defined progressive creation as teaching "that God created many species and after their creation they have varied as the result of

Russia and America (Princeton, NJ: Princeton University Press, 1994), pp. 165–67; Costas
B. Krimbas, "The Evolutionary Worldview of Theodosius Dobzhansky," ibid., pp. 188–90;
Francisco J. Ayala, "Theodosius Dobzhansky: January 25, 1900–December 18, 1975," Na-
tional Academy of Sciences *Biographical Memoirs* 55 (1985): 179. See also Michael Ruse,
Monad to Man: The Concept of Progress in Evolutionary Biology (Cambridge, MA: Harvard
University Press, 1996), pp. 385–401; Dobzhansky's correspondence with John C. Greene in
Greene, *Debating Darwin: Adventures of a Scholar* (Claremont, CA: Regina Books, 1999),
pp. 91–113; and A. Brito da Cunha's reflections on Dobzhansky's growing religiosity in "On
Dobzhansky and His Evolution," *Biology and Philosophy* 13 (1998): 289–300. For a sampling
of Dobzhansky's later views on science and religion, see Theodosins Dobzhansky, "An Essay on
Religion, Death, and Evolution Adaptation," *Zygon: Journal of Religion and Science* 1 (1966):
517–31; Dobzhansky, "Teilhard de Chardin and the Orientation of Evolution," *Zygon: Journal of Religion and Science* 3 (1968): 242–58; Dobzhansky, "Evolution and Man's Conception
of Himself," *Teilhard Review* 5 (1970–71): 65–9; Dobzhansky, "Teilhard and Monod – Two
Conflicting World Views," *Teilhard Review* 8 (1973): 36–40.
28 Russell L. Mixter, "Creation and Evolution," *ASA Monograph* 2 (1950), pp. 16–18; J. F. Cassel
to F. A. Everest, May 5, 1948, ASA Papers (tone down); Cassel to William Wilson, July 4,
1948, Cassel Papers (devil); F. A. Everest to J. F. Cassel, December 28, 1948, ASA Papers
(not fundamentalists). Both the ASA and Cassel Papers are in Special Collections, Buswell
Memorial Library, Wheaton College, Illinois.

mutation and selection so what was once one species has become a number of species, probably as many as are now found in an order or family."[29] Other colleagues pushed beyond orders and families to the entire organic world, and by the early 1960s the ASA had become a hotbed of theistic evolution.

In 1963, fundamentalist dissidents within the ASA broke away to form the Creation Research Society (CRS), dedicated to the defense of a recent special creation and a geologically significant flood. Of the ten founding members, five possessed doctorates in biology: Frank Lewis Marsh (Ph.D. in botany, University of Nebraska), Walter E. Lammerts (Ph.D. in genetics, University of California at Berkeley), William J. Tinkle (Ph.D. in genetics, Ohio State University), John W. Klotz (Ph.D. in genetics, University of Pittsburgh) and Edwin Y. Monsma (Ph.D. in biology, Michigan State University). A sixth member, Duane T. Gish, had earned a Ph.D. degree in biochemistry from Berkeley; a seventh, Wilbert H. Rusch, held a master's degree in biology from the University of Michigan.[30]

The undisputed leader of the group was Lammerts (1904–1996), an outspoken Missouri Lutheran who, unlike many of his colleagues in the society, insisted on the absolute fixity of species.[31] After graduating from the University of California as a Phi Beta Kappa – and a recent convert to Price's flood geology – Lammerts stayed on at Berkeley as a research assistant while he pursued a doctorate in cytogenetics. Following receipt of his doctorate in 1930, making him one of the first strict creationists to earn a Ph.D. in biology, he won a two-year fellowship from the National Research Council for postdoctoral work at the California Institute of Technology. After a brief stint as a research associate back at Berkeley, by which time he had published articles in such journals as the *American Naturalist, Genetics,* and *Cytologia,* he turned to practical plant breeding. From 1940 to 1945 he taught ornamental

---

[29] "The Most Significant Books of the Year," *Eternity* 11 (December 1960): 46; Walter R. Hearn, "Origin of Life," *Journal of the American Scientific Affiliation* 13 (June 1961): 38; R. L. Mixter to V. R. Edman, November 17, 1960, V. Raymond Edman Collection, Box 2, Wheaton College Archives.

[30] W. E. Lammerts to W. J. Tinkle, February 9, 1963; Lammerts to Marsh, May 9, 1963 (secretary); both in the Walter E. Lammerts Papers, Bancroft Library, University of California, Berkeley. Lammerts to Wayne Frair, June 14, 1974 (Germain's), Creation Research Society Papers, Concordia Historical Institute, St. Louis.

[31] George F. Howe, "Walter E. Lammerts," *Creation Research Society Quarterly* 7 (1970–71): 3–4; Lammerts to L. W. Faulstick, September 15, 1962 (grandmother), and Lammerts to D. T. Gish, March 30, 1963 (Berkeley), both in the Lammerts Papers; W. E. Lammerts to G. M. Price, March 27, 1961 (Price), Couperus Papers. See also Walter E. Lammerts, "The Creationist Movement in the United States: A Personal Account," *Journal of Christian Reconstruction* 1 (Summer 1974): 49–63.

horticulture at the University of California, Los Angeles, then left academic
life to help plan the Descanso Gardens in La Cañada.[32]

As Lammerts readily acknowledged, his views on evolution were extreme
even by the standards of creationism. At a time when the strictest creationists
were allowing for – indeed insisting on – considerable evolution within the
created kinds, he held out for "the absolute fixity of species." In his only
departure from this principle, he invoked the miraculous manipulation of
DNA molecules to explain the appearance of new "so-called species" and races
since the original Edenic creation. "As a part of His providential care God may
at various times rearrange the DNA in order to adapt organisms including
man to special conditions," he explained to one Christian biologist, giving the
biblical incident at Babel as an example. "This however is not evolution but a
designed change and would have to be effected rapidly and perfectly in order
to result in a functional organism."[33]

In 1967, Henry M. Morris (b. 1918), coauthor of *The Genesis Flood* (1961)
and the most influential creationist of the group, succeeded Lammerts as
chairman of the CRS board. On most major issues the two creationist ti-
tans generally stood side by side, but they parted company over speciation.
Their most public quarrel erupted in the mid-1970s, when Morris wrote a
letter to the editor of the *Creation Research Society Quarterly* condemning
Lammert's "imaginative reconstruction of the variegated activity of the one
creation and cataclysm clearly described in the Bible." Whereas Lammerts
stuck to "an Ussher-type chronology of only about 7,000 years" and assigned
all of the fossil-bearing rocks to the deluge, Morris allowed for an additional
two or three thousand years of Earth history and placed "the deposits of the
Pleistocene and possibly the Pliocene" after the flood. Morris permitted considerable organic development after the deluge, while Lammerts, to avoid any
evolution at all, attributed post-flood diversity to divine genetic engineering.
Morris viewed such miraculous intervention – amounting, in his opinion, to
a second creation – as theologically suspect because it violated the "economy
of the miraculous in God's orderly world."[34]

In responding to Morris, Lammerts denied that God's genetic tinkering
after the flood could be considered "creations," though it had produced new
forms of plant, animal, and human life. "I will say that if Morris can explain

[32] Interview with Walter E. Lammerts, January 17, 1983; Howe, "Walter E. Lammerts,"
pp. 3–4; Lammerts to Marsh, March 30, 1963 (arguments), and Lammerts to H. C. Doellinger,
November 18, 1963, both in the Lammerts Papers.
[33] Lammerts to V. E. Anderson, May 31, 1965, Lammerts Papers.
[34] Henry M. Morris, letter to the editor, *Creation Research Society Quarterly* 11 (1974–75): 173–5.

how Noah could have been heterozygous for all the distinctive characteristics
of human beings, and then how the races of mankind could have originated by
natural means I will be most happy to accept such an explanation," he replied.
"Evolutionists are hard put to account for this complexity in hundreds of
thousands of years, yet Morris would have this occur in the 5300 or so years
since the Flood, by natural selection of the variation potential which existed
in the survivors of the Flood."[35]

Even Lammert's considerable influence could not stop the majority of
creationist writers from joining Marsh and Morris in calling for extensive
postdiluvian evolution. By the late 1950s, such Young-Earth creationists were
beginning openly to accept *microevolution* (evolution within kinds) as opposed to *macroevolution* (evolution above the level of kinds), terms coined by
the Russian geneticist Iurii Filipchenko (1882–1930) in 1927 and introduced
to the English-speaking world by his most famous student, Dobzhansky. As
one Young Earther proclaimed in 1959, "the creationist can find room for
microevolution or variation, but refuses to accept macroevolution, on the
grounds that it is unscriptural as well as unproved in any form." By the 1990s,
microevolution and natural selection had become standard features of Young-
Earth creationism. When the Alabama State Board of Education in 1995 required that state-approved biology textbooks carry a pasted-in "message"
warning of the controversial nature of evolution, it carefully distinguished
between microevolution, "which can be observed and described as fact," and
macroevolution, such as the development of birds from reptiles, which "has
never been observed and should be considered a theory." Despite the popular
image of creationists being wedded to the fixity of species, no one argued
for more rapid speciation by means of natural selection that those notorious
Darwinian heretics, the creationists.[36]

---

[35] Walter E. Lammerts, letter to the editor, *Creation Research Society Quarterly* 12 (1975–76):
75–7.
[36] Adams, ed., *The Evolution of Theodosius Dobzhansky*, pp. 3, 52; Wilbert H. Rusch, "Darwinism,
Science, and the Bible," in Paul A. Zimmerman, ed., *Darwin, Evolution, and Creation* (St. Louis:
Concordia Publishing House, 1959), p. 33; Marsh, *Variation and Fixity in Nature*, pp. 10–12;
"A Message from the Alabama State Board of Education," reprinted in Numbers, *Darwinism
Comes to America*, p. 10.

CHAPTER SIX

If This Be Heresy

Haeckel's Conversion to Darwinism

Robert J. Richards

Just before Ernst Haeckel's death in 1919, historians began piling on the faggots for a splendid auto-da-fé. Though more people prior to the Great War learned of Darwin's theory through his efforts than through any other source, including Darwin himself, Haeckel has been accused of not preaching orthodox Darwinian doctrine. In 1916, E. S. Russell judged Haeckel's principal theoretical work, *Generelle Morphologie der Organismen*, as "representative not so much of Darwinian as of pre-Darwinian thought."[1] Both Stephen Jay Gould and Peter Bowler endorse this evaluation, seeing as an index of Haeckel's heterodox deviation his use of the biogenetic law that ontogeny recapitulates phylogeny.[2] Michael Ruse, without much analysis, simply proclaims that "Haeckel and friends were not true Darwinians."[3] These historians locate the problem in Haeckel's inclinations toward *Naturphilosophie* and in his adoption of the kind of Romantic attitudes characterizing the earlier biology of Goethe. These charges of heresy assume, of course, that Darwin's own theory harbors no taint of Romanticism and that it consequently remains innocent of the doctrine of recapitulation. I think both assumptions quite mistaken,

---

[1] E. S. Russell, *Form and Function: A Contribution to the History of Animal Morphology* (Chicago: University of Chicago Press, 1982 [1916]), pp. 247–8.

[2] Peter Bowler, *The Non-Darwinian Revolution* (Baltimore: Johns Hopkins University Press, 1988), p. 83–4.

[3] Michael Ruse, *Monad to Man: The Concept of Progress in Evolutionary Biology* (Cambridge, MA: Harvard University Press, 1996), p. 181.

Figure 6.1. Ernst Haeckel (right) and his assistant and student Nikolai Miklucho-Maclay.
Photo about 1866, courtesy of Haeckel Haus, Jena.

and have so argued.[4] But against the charge of heresy, one can bring a more
direct and authoritative voice – Darwin himself.

In 1863, Haeckel made bold to send Darwin his recently published two-
volume monograph on radiolarians – one-celled aquatic animals that secrete
a skeleton of silica. The first volume examined in minute detail the biology

---

[4] I have argued that the recapitulational thesis forms the heart of Darwin's own theory of evolution and that Darwin's conception of nature derived from Romantic sources. See Robert
J. Richards, *The Meaning of Evolution: The Morphological Construction and Ideological Re-
construction of Darwin's Theory* (Chicago: University of Chicago Press, 1992), pp. 91–166;
and Richards, "Darwin's Romantic Biology, the Foundation of his Evolutionary Ethics," in
Jane Maienschein and Michael Ruse, eds., *Biology and the Foundation of Ethics* (Cambridge:
Cambridge University Press, 1999), pp. 113–53. See also the exchange: Peter Bowler, "A Bridge
Too Far," *Biology and Philosophy* 8 (1993): 98–102; and Robert J. Richards, "Ideology and the
History of Science," *Biology and Philosophy* 8 (1993): 103–8.

of these creatures and argued that Darwin's theory made their relationships comprehensible. The second volume contained extraordinary copper-plate etchings depicting the quite unusual geometry of these animals. Darwin immediately replied to this previously unknown zoologist that the volumes "were the most magnificent works which I have ever seen, & I am proud to possess a copy from the author."[5] Emboldened by his own initiative in contacting the famous naturalist, Haeckel, a few days later, sent Darwin a newspaper clipping that described a meeting of the Society of German Naturalists and Physicians at Stettin, which had occurred during the previous autumn. The article gave an extended and laudatory account of Haeckel's lecture defending Darwin's theory.[6] Darwin quickly responded in his second letter: "I am delighted that so distinguished a naturalist should confirm & expound my views; and I can clearly see that you are one of the few who clearly understands Natural Selection."[7] Darwin thus judged Haeckel a true disciple, "one of the few who clearly understands Natural Selection" and one whose research ability and aesthetic sense lent considerable weight to the new evolutionary theory. Darwin thus stands as a witness against later scholars who wish to cast Haeckel from the camp of the orthodox.

Of course, contemporary historians might argue that Darwin did not understand the full scope of Haeckel's own biological ideas and that had his German been better he would have detected deviant tendencies in the work of his new disciple. In this chapter, I wish to provide further evidence that Darwin was not mistaken in his original evaluation. The full argument for this position must be postponed, but a good start can, I believe, be made by following in measured step the road Haeckel took to Damascus.

EARLY STUDENT YEARS

Ernst Heinrich Haeckel was born February 16, 1834, in Potsdam, where his father Karl (1781–1871), a jurist, served as privy counselor to the Prussian

---

[5] Charles Darwin to Ernst Haeckel, March 3, 1864, in the Correspondence of Ernst Haeckel, in the Haeckel Papers, Institut für Geschichte der Medizin und der Naturwissenschaften, Ernst-Haeckel-Haus, Friedrich-Schiller-Universität, Jena. Hereinafter I will refer to this as "Haeckel Correspondence, Haeckel-Haus, Jena."

[6] *Stettiner Zeitung* (nr. 439), Sept. 20, 1863. The author began: "The first speaker [Haeckel] stepped up to the podium and delivered to rapt attention a lecture on Darwin's theory of creation. The lecture captivated the auditorium because of its illuminatingly clear presentation and extremely elegant form." The author then gave an extensive précis of the contents of the entire lecture. He concluded by reporting that "a huge applause followed this exciting lecture."

[7] Charles Darwin to Ernst Haeckel, March 9, 1864, Haeckel Correspondence, Haeckel-Haus, Jena.

court. His mother Charlotte (née Sethe,1767–1855) nurtured him on classic German poetry, especially that of her favorite, Friedrich Schiller, while his father discussed with him the nature-philosophy of Johann Wolfgang von Goethe and the religious views of Friedrich Schleiermacher, who had been an intimate of the family, especially of Haeckel's aunt Bertha. Karl Haeckel had a keen interest in geology and foreign vistas, which undoubtedly led his son to treasure the travel literature of Alexander von Humboldt and Charles Darwin, which the boy devoured, and later to yearn for a life of adventure in exotic lands. Haeckel's judicial heritage may also have fostered a lingering impulse to bring legal clarity, through the promulgation of numerous laws, into what he perceived as ill-ordered biological disciplines.

Medical School at Würzburg

Though Haeckel had harbored the desire to study botany at university, he acceded to his father's wishes and, in August 1852, enrolled in the medical school at Würzburg. The university at that time had probably the best medical faculty in Germany. Students – some six hundred in 1852 – came from everywhere to study with such luminaries as Albert von Kölliker (1817–1906) and Rudolf Virchow (1821–1902). Kölliker taught histology and introduced Haeckel to what would quickly flower into a sweet delight – at least for one so disposed – namely, microscopic study; and Kölliker's just-published *Handbuch der Gewebelehre* (1852) became his *vade mecum*. But the star of the faculty was Virchow, whose history of political engagement excited a frisson of danger in the active imaginations of his students. His ideas concerning the cellular basis of life and disease proved just as radical as his politics had been; and his reputation for deep research and academic controversy ensured that his lectures would be jammed. His electrifying talent as a scientist indeed drew Haeckel to his classes, but his insulated and cool personality kept the two from becoming close – quite in contrast to Haeckel's relationship with Kölliker, with whom he would strongly disagree intellectually but with whom he would remain on warm personal terms. Virchow and Haeckel would later interact in proper professional ways – until, that is, the famous senior scientist began preaching the dangers of evolutionary theory for untutored minds. In 1877, Virchow recommended to his colleagues that they not press for evolutionary theory to be taught in the German middle and lower schools, since, as he argued, it lacked scientific evidence, was an affront to religion, and smoothed the way to socialism. Haeckel's sulfuric reaction to this admonition undoubtedly released a force building since his student days.[8]

Haeckel did not take naturally to the idea of medical school and its likely consequence, clinical practice. Two lines, though, seemed to have kept him tethered to medicine: a tempered passion for the kind of fundamental science that he experienced with Kölliker and Virchow, and a strategy for utilizing medicine to achieve the scientific vocation that he envisioned from his reading of Humboldt. Under the affable tutelage of Kölliker, he grew to love precise work in histology, especially since he had a talent with the microscope. He could simultaneously peer with one eye through the lens and with the other draw in exquisite detail the minute structures of tissues. "Vivant cellulae! Vivat Microscopia!" he exulted to his father at Christmas in 1853. But it was Virchow's lectures during his second year that confirmed him in a resolve, made to his father, to stick with medicine. He provided his father a description of the arresting experience:

Virchow's lecture is rather difficult, but extraordinarily beautiful. I have never before seen such a pregnant concision, a compressed power, a tight consistency, a sharp logic, and yet the most insightful descriptions and compelling liveliness as are here united in lectures. Though, if one does not bring to the lectures an intense concentration and a good philosophical and general culture, it is very difficult to follow him and to get a hold of the thread that he so beautifully draws through everything; a clear understanding will be taxed considerably by a mass of dark, quickly moving expressions, learned allusions, and a large use of foreign terms, which are often very superfluous.[9]

Kölliker and Virchow, by the force of their personalities, made deep impressions on the fledgling researcher. They taught him the value of bold hypothesis and precise empirical research. But two other German scientists – by reason of their philosophical and aesthetic views – were to have a more profound impact on Haeckel's conception of nature and his future adoption of

---

[8] German government officials were preparing for the lower schools' pedagogical reforms in which natural science would play a more important role than it had. Virchow cautioned his colleagues, at the meeting of German Natural Scientists and Physicians at Munich in 1877, not to insist that evolutionary theory become part of the curriculum. He thought only the secure facts of science ought to be represented, and evolutionary theory had no real empirical support – especially those aspects that Haeckel contended for, namely, spontaneous generation and the descent of man. See Virchow's talk given at the Versammlung deutscher Naturforscher and Ärtze at Munich, September 1877, reprinted as *Freiheit der Wissenschaft im modernen Staat* (Berlin: Wiegandt, Hempel & Parey, 1877). See also Haeckel's reply in his *Freie Wissenschaft und freie Lehre: Eine Entgegnung auf Rudolf Virchow's Münchener Rede* (Stuttgart: Schweizerbart'sche Verlagshandlung, 1878).

[9] Ernst Haeckel to his parents, November 16, 1853, in Heinrich Schmidt, ed., *Entwicklungs-geschichete einer Jugend: Briefe an die Eltern, 1852–1856* (Leipzig: Koehler, 1921), p. 80.

evolutionary theory. These were Alexander von Humboldt (1769–1859) and
Johann Wolfgang von Goethe (1749–1832).

The Aesthetic Science of Humboldt and Goethe

In his *Voyage aux Régions Equinoxiales du Nouveau Continent, fait en 1799–
1804* (Travel to the equinoctial regions of the new continent, made from
1799–1804, published 1807–34), in his *Anschichten der Natur* (Views of nature,
1849), and especially in his famous *Kosmos* (Cosmos, 1845–62), Humboldt
attempted to formulate and plait together a great many empirical laws – those
characterizing astronomy, chemistry, physics, geology, botany, and zoology.
He believed that the principles of those several disciplines touching on the
phenomena of life all harmoniously articulated with one another and thus
demonstrated that "a common, lawful, and eternal bond runs through all
of living nature."[10] The task of the natural scientist, then, was to reveal this
harmony of laws producing a unified whole, to work through the vast and
wondrous diversity of nature to discover the underlying forms. The harmony
of nature – a *cosmos*, according to Humboldt – was discovered to both reason
and poetic imagination. He himself proposed many quantitative principles of
plant morphology and biogeography. But he was equally insistent about the
necessity of cultivating the aesthetic aspects of nature, since aesthetic judgment
was no less important for human understanding than mechanistic determination. "Descriptions of nature," Humboldt observed in a Kantian vein,

can be sharply delimited and scientifically exact, without being evacuated of the viv-
ifying breath of imagination. The poetic character must derive from the intuited
connection between the sensuous and the intellectual, from the feeling of the vastness,
and of the mutual limitation and unity of living nature.[11]

This same basic premise, that scientific judgments and aesthetic judgments
about living nature have the same structure and aim – that they deliver to
comprehension the unity and diversity of nature, but portend the sublime –
this premise was of Kantian origin but likely of more immediate Goethean
derivation. It had been a subject of some conversation between Goethe and
Humboldt during the many years of their friendship.

Goethe anchored the principle of complementarity of scientific and aesthetic judgment in his metaphysical monism, a conception Haeckel himself

---

[10] Alexander von Humboldt, *Kosmos. Entwurf einer physischen Weltbeschreibung*, 5 vols.
(Stuttgart: Gotta'schen Buchhandlung, 1845–62), vol. 1, p. 9.
[11] Ibid, vol. 2, p. 74.

would adopt. Goethe, following Spinoza, conceived of nature as harboring *adequate ideas*, archetypes that the naturalist had to recover in order to articulate nature in scientific law and theory, and that the artist had to comprehend in order to render natural beauty in painting and poetry.[12] Haeckel's consumption of great quantities of Humboldt and Goethe during his medical school years caused his own ideas to pulse with their conceptions of science and art.

The Research Ideal

Goethean and Humboldtean ideas fueled Haeckel's own natural propensities. During his medical school days, he was hardly a solitary figure. He had good friends among his classmates, with whom he learned to lift a pint, at least on occasion; he also had several acquaintances among the faculty.[13] But in those moments of adolescent's deep reflections and inevitable anxieties, he found great consolation in the Romantics' traditional resources – nature and poetry. After dinner, with a friend or alone, he would often steal out into the countryside to savor the delights of nature settling into evening. Or in the twilight of his darkening room, he would light a candle and pull down his Schiller, Goethe, or perhaps read from a translation of Shakespeare – a favorite of the Romantics.

Though he often felt he had two souls dwelling in one breast – that of the "loving man," who feels deeply and kindles his passions with nature and poetry, and that of the "scientific man," who splashes cold reason on the emotions in order to achieve objective understanding – he yet conceived of a way to temper these disjoint inclinations. This was through the Humboldtian vision of the researcher who works in exotic lands and occasionally attends to the medical needs of the natives. He used this image to fortify his efforts at medicine, which he ever hated. It was an adolescent dream, but one that, remarkably, would materialize in a few years.

[12] I have discussed Goethe's scientific and aesthetic principles in my *The Romantic Conception of Life: Science and Poetry in the Age of Goethe* (Chicago: University of Chicago Press, 2002), Chapters 10 and 11.
[13] Haeckel was a bit of a loner but not without good and close friends, who made some concerted efforts to loosen him up. On one occasion, they enticed him to attend a masked ball held in Würzburg. When he got there, he was astonished to have a mysterious young woman – at the time he knew only two women, both wives of professors! – come up and chide him for not socializing more. Delighted, he asked her to write her name on a slip of paper. "Mysterious" was the name she wrote, and then vanished. Haeckel suspected that his friends had put her up to it. This and like experiences brought him to a certain resolution: the next year he took dancing lessons. See his letters to his parents (of March 20, 1845 and November 19, 1855, in Schmidt, ed., *Briefe an die Eltern*, pp. 107 and 167.

Perhaps no experience confirmed Haeckel in his goal of biological (as opposed to medical) research more than his new relation with the most famous physiologist and zoologist of his day, Johannes Müller. In spring of 1854, Haeckel decided to take his summer term in Berlin. Away from provincial Würzburg, he would indulge himself in this "metropolis of intellect" and, of course, visit with his parents and relatives. He would also have opportunity to study with the renowned Müller.[14]

During the summer term at Berlin, Haeckel attended Müller's lectures on comparative anatomy and physiology. The decisive experience with Müller, though, came during the summer vacation. At the end of August 1854, Haeckel and his friend Adolph de la Valette St. George decided to travel to Helgoland (two islands in the North Sea, west of Schleswig-Holstein). They planned to meet other student friends there for seaweed collecting and rather desultory anatomical study – all to be refreshed by a good deal of sea bathing. Most likely, Müller's stories of collecting off the islands, along with other tourist delights, inspired them to go. On the way, they passed through the port city of Hamburg, whose shops carried exotic wares from all over the globe and whose streets could hardly contain the crowds of sailors, tourists, peddlers, and citizens of all stations and dress. The harbor itself displayed to the entranced students a tangled forest of masts and rigging from ships that plied the seas of the world. After a harrowing passage on a new three-masted iron steamer during a great gale, Haeckel and la Valette disembarked on the principal island of Helgoland in the late afternoon of August 17. They settled into a routine of sea bathing at 6:00 A.M. and collecting and dissecting during the rest of the day. It was a revealing experience for Haeckel, as he indicated to his parents: "You cannot believe what new things I see and learn here every day; it exceeds by far my most exaggerated expectations and hopes. Everything that I studied for years in books, I see here suddenly with my own eyes, as if I were cast under a spell, and each hour, which brings me surprises and instruction, prepares wonderful memories for the future."[15]

Rather unexpectedly, Johannes Müller and his son Max arrived in Helgoland for two weeks of research on echinoderms (starfish, sea urchins, etc.). Müller immediately invited Haeckel and la Valette to accompany his son and him on their fishing and research expeditions. The friendship of his revered teacher and the marvel of the invertebrates they brought up for study each day irrevocably altered the course of Haeckel's research interests, from botany to marine invertebrate zoology, a transition sealed with

---

[14] Ernst Haeckel to his parents, March 25, 1854, ibid., p. 109.
[15] Ernst Haeckel to his parents, August 30, 1854, ibid., p. 122.

the publication the next year of his maiden research article in Müller's *Archiv*.[16]

Haeckel extended his stay in Berlin through the winter semester of 1854–55 but returned to Würzburg the following spring. He spent the summer term of 1855 in clinical training and in the fall would commence the actual treatment of patients. During the summer, though, he also found time to take a short course in the dissection of invertebrates offered by the *Privatdozenten* Franz Leydig (1821–1905) and Carl Gegenbaur (1826–1903), both of whom worked with Kölliker. Haeckel's clinical experience was mostly confined to the poor and destitute of Würzburg, and the cases with which he dealt – in children, for example, horrible worms, rickets, scrofula, and eye diseases – did little to stimulate his appetite for the practice of medicine. The only part he really enjoyed was the postmortem anatomies, of which there seemed to be no short supply. His salvation during this period lay in the tutelage of Virchow, who encouraged the young student in pathological anatomy. Virchow oversaw Haeckel's next two publications, which embroiled him in a controversy with his mentor's opponents.[17] "But how sweet to be attacked in defense of Virchow," he wrote to his parents.[18] After a successful competitive anatomy exam, Haeckel became Virchow's assistant for the summer of 1856, and harbored the hope that the great man would take him along in the autumn to the University of Berlin, to which the renowned scientist had been called. But during that summer, Haeckel began again to despise the clinical practice of medicine and longed to be able to pursue what he thought his true vocation – biological research. Moreover, though his relationship with Virchow was cordial, the cool and reserved character of the professor ill complemented the passionate and volatile nature of the student.

After the tedious summer weeks of clinical work, Haeckel was invited by Kölliker to travel with him to Nice for collecting and anatomical study of invertebrates. Haeckel rejoiced at the opportunity, made good with the help of some 150 Reich's dollars from his father. On the beautiful French Riviera, the company met Johannes Müller, and the whole experience convinced the young scientist that he had entered paradise. But the bliss of biology gave way again to dreaded medicine, and in the winter semester of 1856–57, Haeckel

---

[16] Ernst Haeckel, "Über die Eier der Scomberesoces," *Archiv für Anatomie und Physiologie* 22 (1855): 23–32.

[17] Ernst Haeckel, "Zwei medizinische Abhandlungen aus Würzburg: I. Über die Beziehungen des Typhus zur Tuberkulose; II. Fibroid des Uterus," *Wiener medizinische Wochenschrift* 6 (1856): 1–5, 17–20, 97–101.

[18] Ernst Haeckel to his parents, June 8, 1856, in Schmidt, ed., *Briefe an die Eltern*, pp. 184–8.

retreated to Berlin to prepare his medical dissertation, which he wrote under the guidance of Leydig. His study was on the histology of river crabs (*De telis quisbusdam Astaci fluviantilis*),[19] a subject of conveniently ambiguous disciplinary direction. He received his doctorate in March 1857 and then felt compelled to spend the summer in Vienna engaged in further clinical study to prepare for the state medical exam, which, after more anxious preparation in Berlin during the winter semester, he passed the following March.

During his medical education, Haeckel became ever more passionate about his vocation – not that of a physician, but that of a biological researcher, one whose ideal was formed in the exacting microscopical work done under the guidance of Kölliker and Virchow but whose deeply rooted inclinations drew him toward the kind of science practiced by Humboldt and Goethe.

HABILITATION AND ENGAGEMENT

After passing his state medical examinations, Haeckel laid plans for the prosecution of his true vocation, research science. He arranged with Johannes Müller to conduct his habilitation study at Berlin – the habilitation, with its required monograph, being a prerequisite for an academic position. During this period, though, Müller suffered from the deepest of depressions, which led him to the ultimate solution. He took his own life with an overdose of opium – at least that was what Haeckel suspected.[20] Haeckel was devastated, not simply because of a lost opportunity, but because he truly revered and loved the man.

Haeckel's academic ambitions brightened when another Müller protégé, Carl Gegenbaur, his friend from Würzburg, invited him to visit Jena, where Gegenbaur had become ordinary professor of anatomy in the medical faculty.[21] During the visit in May 1858, Gegenbaur offered intimations of

---

[19] Ernst Haeckel, *De telis quibusdam Astaci fluviatilis* (Berlin: Schade, 1857).

[20] While in his late sixties, Haeckel became enamored of a beautiful young woman over thirty years his junior, Frieda von Uslar-Gleichen. He thought the relationship doomed, and in their voluminous correspondence he would often pour out his despair. In his letter to her of January 11, 1900, he mentioned that he often thought of suicide and that his "great, highly revered master, Johannes Müller, ended his nervous condition (accompanied by sleeplessness) with morphine in April 1858." (The correspondence of Ernst Haeckel and Frieda von Uslar-Gleichen is held in the manuscript room of the library of the Preuschische Kulturbesitzt, Berlin.) Gottfried Koller judges otherwise, suggesting that any overdose of morphine would have been accidental. See his *Das Leben des Biologen Johannes Müller, 1801–1858* (Stuttgart: Wissenschaftliche Verlagsgesellschaft, 1958), pp. 234–36.

[21] Gegenbaur initially came to Jena as extraordinary professor (roughly the equivalent of an American associate professor) in 1855. In 1851 he had met Johannes Müller, who persuaded

support, and more straightforwardly asked Haeckel if he would care to travel to Messina in October with him. To Haeckel it seemed a dream materialized, and he quickly said yes. The dream began to dissolve, however, when Gegenbaur and Moritz Seebeck (1805–1884), the curator of the university, took him aside to offer the advice of wisdom and age – that he should not even think about marriage lest his scientific career sink before being properly launched. That evening, with obviously troubled conscience, Haeckel sat down to write of this conversation to Anna Sethe, his first cousin and the woman to whom he had become secretly engaged two days after Müller's burial.[22]

Haeckel had first met his cousin at the wedding of his brother Karl and Anna's sister Hermine. In his diary for September 21, 1852, when he was eighteen and she seventeen, he penned: "Celebration at Karl's wedding. Anna Sethe as an elf! Dancing. I knew how but couldn't dance and sat (as usual when others are having fun) in a melancholy mood by myself in the back of the room."[23] Haeckel would see Anna from time to time at various family gatherings. In 1856, she came with Haeckel's parents to visit him in Würzburg. After the death of her father, she and her mother moved to Berlin in 1857, during the time Haeckel was there working on his dissertation. Through the next year their relationship ripened, and in precipitous passion at the time of Müller's death, he asked her to marry him. It was only two months later that Gegenbaur and Seebeck offered their peremptory advice, which was often repeated by friends and relatives to whom he revealed his secret.[24]

The difficulties of managing both marriage and a career – a career that had not even really begun – agitated Haeckel through the summer of 1858 and beyond. But at the same time he came to perceive Anna as the lodestar of his life – even more, as an all-consuming love that gave meaning to his work and, it is no exaggeration, to the universe. She was in many ways the young,

---

him to do research in Helgoland; the next year, with Heinrich Müller and Kölliker, Gegenbaur traveled to Messina, where he became confirmed in his interest in marine invertebrates. He habilitated at the end of the winter term of 1853–1854 with a monograph on generational alteration and reproduction in *Medusae* and polyps. See Carl Gegenbaur, *Erlebtes und Erstrebtes* (Leibzig: Engelmann, 1901), pp. 57–64, 87; and Georg Uschmann, *Geschichte der Zoologie under der zoologischen Anstalten in Jena 1779–1919* (Jena: Fischer, 1959), pp. 28–9.

[22] Ernst Haeckel to Anna Sethe, May 25, 1858, in Heinrich Schmidt, ed., *Himmelhoch Jauchzend: Erinnerungen und Briefe der Liebe* (Dresden: Reissner, 1927), p. 19. This volume contains Haeckel's letters to Anna from spring 1858 to fall 1862.

[23] The passage from Haeckel's diary is quoted by Schmidt in the introduction to Haeckel's letters to Anna, ibid., p. 6.

[24] Haeckel mentioned to Anna these several warnings. See his letters to Anna of April 9 and September 27, 1858, in ibid., pp. 64–5, 76.

long-haired, blond, blue-eyed scientist's female double, either in blood or in
his own imagination, as his description for a friend suggests:

A true German child of the forest, with blue eyes and blond hair and a lively natural
intelligence, a clear understanding, and a budding imagination. She puts no stock in
the so-called higher and finer world, for which I hold her even higher since she was
brought up in it. She is rather a completely unspoiled, pure, natural person.[25]

Haeckel's letters to Anna over the period of their courtship express three
intertwined themes: his love for her; his hopes of landing a professorship,
which would allow them to marry; and his exuberant and irrepressible attachment to nature, an emotion that at times seemed to rival his love for her.
But through this period, the latter themes gradually become submerged in
an overflow of desire for Anna. "How our souls have already so closely and
strongly grown together," he exclaimed to her in August, "so that absolutely
nothing can separate them and so that every thought and every action are
able to be realized only with and in the 'other ego.'" He thought of her love
as a kind of salvation, a lifeline that would pull him back from the dark abyss
of materialism toward which he felt himself dragged by his science. "When I
press through from this gloomy, hopeless realm of reason to the light of hope
and belief – which remains yet a puzzle to me – it will only be through your
love, my best, only Anna."[26]

Their growing love pressed them to reveal officially what by midsummer
most of their friends knew already; and so on September 14, 1858, in Anna's
new family home in Heringsdorf (north of Berlin on the Baltic Sea), they
announced and celebrated their engagement. Two weeks later, Haeckel wrote
to his fiancée from Berlin, recalling with febrile delight their Sunday morning
walk on that festive day.

My gay, frisky roe trotted by my side, happy and free over rocks and roots, slipping
through thorns and thickets. [They sat down on the green-moss bank] and your sighing
breath, your warm cheek on mine announced to me at every blissful second that sweet
unspeakable happiness that I held in my arms, close and sure, so that I might never,
never lose it. Then we lay on my good old plaid, placed on the natural bed of the forest,
upholstered with dry beech leaves, sloping down on the side, at the foot of two old
boughs, and we peered through the thousand smaller and larger holes carved out for
us between the round, green leaves up into the deep blue cloudless sky, whose bright
sun so wonderfully shown on the happy pair as if it rejoiced with them. O Anna, those
were moments I will never, never forget, moments of the greatest human happiness,
the most happy because the individual himself is completely forgotten; he removes

---

[25] Ernst Haeckel to a friend, September 14, 1858, ibid., p. 67.
[26] Ernst Haeckel to Anna Sethe, August 22, 1858, ibid., p. 54.

himself purely and completely from the dirty, spotted veil of a suffering personality in which he is wrapped, and lifts himself up and beyond into a full and pure intuition of the other in the joy of an absolute giving to the other. One forgets heaven and heart, past and future, and lives purely and completely in the present. Here Faust himself could exclaim, 'Tarry a while, you are so beautiful,' so he might secure the moment which sadly only too quickly dissolves.[27]

During the August prior to their engagement, Haeckel had traveled again to Jena, invited by Gegenbaur for the celebration of the three-hundredth anniversary of the university. During this visit, his new mentor mentioned that he would probably not travel to Italy, and so their planned trip together would be cancelled. Haeckel decided he had to make the trip nonetheless, even if he had to go it alone. It would be an excursion not simply to secure a subject for his *Habilitationschrift*, but also one of *Bildung*, of intellectual and personal formation. He planned to spend the spring of 1859 in Florence and Rome, studying art; in summer to travel to Naples, where he would begin his marine research; and to finish in Palermo and Messina in winter. He expected his travel would "reform and give rebirth to my whole outlook on life."[28] Both Haeckel's itinerary and his sentiment echo Goethe's, as described in the poet's *Italienische Reise* (Italian journey, 1816–29).

LOVE AND RESEARCH IN ITALY

On January 28, 1859, Haeckel left Berlin and traveled back to Würzburg to collect materials and equipment. He then went on to Luzerne, Genoa, and on February 6 he arrived in the artistic heart of Italy, Florence. But for Haeckel, the heart beat dull and weak. He intended to study and copy the masterpieces that seemed to hang on every wall of the city. But quickly he grew weary of the incessantly repeating themes – like Noah's Flood, biblical stories gushed from every wall. And then there were the countless Madonnas: Mary as a child, the Annunciation, the Birth, the Domestic model, the Grieving Mother, and now as a French woman, an Italian, a German, or a Spaniard, and each depicted in the habits of every century. The art was too religious, too Catholic, too much for Haeckel's north German sensibilities. He traveled to Pisa for relief in mid-February, but again he was surrounded by artful Virgins.

The Eternal City, which he reached on February 23, seemed even more heavily caked with the cloying oils of southern religious sentiment. But worse yet, almost daily the streets of the ancient city would be choked off with religious

---

[27] Ernst Haeckel to Anna Sethe, September 26, 1858, ibid., pp. 72–3.
[28] Ernst Haeckel to Anna Sethe, August 22, 1858, p. 56.

processions in celebration of one of the innumerable saints of the Roman calendar. He would see cardinals from this or that cathedral riding in their gilded coaches and displaying to the poor of the city their scarlet robes be-decked with jewels. He wrote to Anna that "had I not already during the last years – through a study of nature, pressing into her depths and finest parts – discarded the Christianity of the theologians, here in Rome I would surely become a pagan."[29]

Beneath the façade of the citadel ruled by "the Pope with his band of Christian barbarians," Haeckel found the ancient city of Virgil, Horace, and Cicero. In the moonlight, he would walk through the ruins of that ghostly civilization and conjure up the shade of Goethe, who had passed along the same way during his own Italian journey three-quarters of a century before.[30] But unlike Goethe, who could delight in the pomp of Catholicism, the craft of the Jesuits, and the decadence of the streets (especially its women of easy instruction), Haeckel felt suffocated. He left Rome on March 28 and traveled to Naples, where he had to get to the chief business that had brought him to Italy, biological research.

Naples was no joy. He had barely adequate accommodations, with constant noise from the streets. In the spring, the weather was awful – frequent rain interrupted by oppressive heat, and the unremitting winds of the sirocco out of North Africa. Nor did the Neapolitans elevate his estimation of humankind: "The dishonesty, superficiality, thoughtlessness, the swindling selfishness overreaches all the usual bounds and for a true German this is all doubly painful."[31] Anna diagnosed his unhappiness in Naples as a consequence of his loss of religious faith. Haeckel agreed with this analysis, but protested that even if he were ten times as unhappy, he could "never again accept an arbitrary dogma." "The fruit of the tree of knowledge," he wrote to his Eve, "is worth the loss of Paradise."[32]

Despite his discomforts, Haeckel settled for almost six months in Naples, until mid-September. After he had obtained a modestly regular and reliable supply of catch from the local fishermen, he spent most of his day – roughly from 9:00 A.M. to 5:30 P.M. – in examining and describing the various invertebrates that slithered across his table. But he had no direction in his research, and many creatures easily slipped through the gaps in his knowledge. He began

[29] Ernst Haeckel to Anna Sethe, February 28, 1859, in Ernst Haeckel, *Italienfahrt: Briefe an die Braut, 1859–1860*, ed. Heinrich Schmidt (Leipzig: Koehler, 1921), p. 8.
[30] Haeckel described his experiences in Rome to Anna in letters of February 28, March 1, and March 15, 1859, ibid., pp. 8–9, 14.
[31] Ernst Haeckel to Anna Sethe, April 18, 1859, ibid., p. 28.
[32] Ernst Haeckel to Anna Sethe, May 29, 1859, ibid., p. 65.

to despair of ever becoming master of the field and of discovering something significant, which did not bode fair for attaining an academic position and marrying Anna. Despite her constant efforts to cheer him, the lines of *Faust* came liquid to his pen: "I am plagued by no scruple or doubt, nor do I fear hell or the devil; yet all joy has been ripped from me, and I do not imagine I can know anything aright or teach anything to better men and convert them."[33]

FRIENDSHIP WITH ALLMERS AND TEMPTATIONS OF THE BOHEMIAN LIFE

On June 17, no longer able to stomach the city, Haeckel took palette and easel, and slipped across the bay of Naples to the beautiful island of Ischia. Under a sunny sky and surrounded by mountains and small forests, ripe for sketching or botanizing, Haeckel's mood shifted to contentment and then to something like happiness. But what made the trip more than a mood elevator was his meeting there with the poet and painter Hermann Allmers (1821–1902), who would become his lifelong friend. Haeckel found in Allmers the odd complement. Allmers was fourteen years older, gnomelike in appearance, and possessed of a "colossal Bedouin nose"[34] – the opposite of the tall, golden, and strikingly handsome young scientist. The magnetic pull reached down to the souls of each, as Haeckel reported to Anna:

Allmers is above all a poet. He sees the whole of life, with all its light and shadowy sides, only from the beautiful, misty perspective of poetry, and so constitutes in this idealism a stark contrast to my natural-scientific realism, which strives to discard this misty, yet so very beautiful, gown and to view reality generally in its naked truth.[35]

These complements of talent and attitude – running over a deeper sexual feeling – supplemented, however, the more repressed inclinations of each: Allmers could botanize with exactitude, and Haeckel often very happily would lose himself "in the misty distances of a dreamy poetry."[36]

Haeckel's friendship with Allmers slowly drew him away from steady work in biology. In August they sailed to Capri, where they would spend the month

---

[33] Ernst Haeckel to Anna Sethe, May 9, 1859, ibid., pp. 49–50.

[34] Haeckel, in a very friendly fashion, recalls his time with Allmers on Ischia, a time when he looked forward to seeing that "collosal Bedouin nose" coming around a mound of rocks. See Ernst Haeckel to Hermann Allmers, May 14, 1860, in Rudolph Koop, ed., *Haeckel und Allmers: Die Geschichte einer Freundschaft in Briefen der Freunde* (Bremen: Arthur Geist Verlag, 1941), p. 46.

[35] Ernst Haeckel to Anna Sethe, August 1, 1859, in Haeckel, *Briefe an die Braut*, p. 79.

[36] Ibid., p. 80.

leading a bohemian life of wandering through the countryside, bathing, and painting. Capri seemed to Haeckel the realization of the dreams of his youth, dreams arising out of his reading of *Robinson Crusoe* and the accounts of Humboldt's and Darwin's travels, even if this Italian island melted into a glow the hardships described in those earlier works. With the beauty of the island, the companionship of the other artists there, and the deepening friendship of Allmers, Haeckel was tempted to abandon his thus far fruitless research and spend his days in landscape painting – his great delight – and his nights in dancing the tarantella, as he had on the night of their departure from Capri. What restrained this possibility was that Haeckel recognized his talent with watercolors was somewhat less than his aspiration, and, of course, it was obvious that the life of the bohemian did not pay very well, certainly not enough to support a wife, his Anna, to whom he felt ever closer.

Haeckel's itinerary now dictated that he leave for Messina, the Sicilian city where his revered teacher Müller had spent so many profitable days. Forty-eight hours after they returned from Capri, Haeckel and Allmers arrived in Messina, on September 10, 1859. They spent five weeks together traveling around the island by ship, wagon, mule, and foot. Compared to Capri, which remained his "Italian Paradise," Sicily was disappointing in its quite ordinary flora and fauna. The forest had almost disappeared, and the cities had little to recommend them. Only ancient ruins offered some interest to the travelers. Haeckel found the Sicilians more to his liking than the Neapolitans, though only by a breath. "The Sicilians," he wrote Anna, "even if they are not comparably so depraved, so bereft of all virtue and honor as the completely bovine Neapolitans, they are, nonetheless, such a miserable group that a sensitive German conscience could never be reconciled to their superficial considerations and aspirations."[37]

In mid-October, Allmers had to leave, and Haeckel at last turned to work. He justified to Anna, and more especially to himself, his time in Italy thus far as necessary for the development of his mind and character, for the deepening of his appreciation of natural beauty. It was the sort of *Bildung* experienced by Goethe on his own Italian journey, and Haeckel hoped for some comparable outcome.[38]

The flood of creatures that welled up in the seas around Messina, "the Eldorado of zoology," he called it, drove Haeckel to despair of seizing and reducing to actuality that great wealth of possibilities. Not only was he

[37] Ernst Haeckel to Anna Sethe, October 16, 1859, in Haeckel, *Briefe an die Braut*, p. 112.
[38] Ernst Haeckel to Anna Sethe, October 21, 1859, ibid., pp. 116–17.

delivered of unusual species and genera, he encountered whole families, orders, and classes never before described, beautiful and astonishing animals – siphonophores, petropods, heteropods, radiolaria, medusas, and so on. As the mountain buried him in its avalanche of goods, Haeckel pulled back into thoughts of the artist's existence, which promised "a rich, creative, colorful life of imagination, while that of the scientist offers a sober, cold, anatomical effort of reason that always soon leads to negation and skeptical dissolution, a reason that is oriented to a possible understanding of natural wonder that we can never comprehend."[39] What kept him from casting off his plan – which now desiccated into that of "a repressed professor who in Jena or Freiburg or Tübingen or Königsberg or in some other small, petty university, every semester must take his one-and-a-half to three students and 'here and there, back and forth, lead them by the nose' " – what constrained him on that gloomy professorial path was the image of the bright presence of Anna, who awaited at the end. Haeckel's despair at this juncture formed the negative image of his recent, glorious experience with Allmers and the desire for the distant Anna. But the bitter taste of research would quickly turn sweet as the topic for his *Habilitationschrift* began to congeal.

RADIOLARIA

At the end of November, with just a few months left for his research in Italy, Haeckel finally decided to focus on just one group of animals, the almost unknown radiolaria – a large class of one-celled marine organisms that secreted unusual skeletons of silica.[40] When he had traveled in the summer of 1856 with

---

[39] Ernst Haeckel to Anna Sethe (21 October 1859), ibid., pp. 118–19.

[40] In 1836 and 1837, Christian Gottfried Ehrenberg (1795–1876) had described conglomerates of fossil protozoa, among which were, apparently, some radiolaria and perhaps *Acantharia*, distinguishable by the chemical composition of their skeletons, which in the fossilized state he described respectively as silica and flint. (Nonfossilized radiolaria and *Acantharia* skeletons we now know to be composed, respectively, of silica and strontium sulfate.) These remains were similar, he maintained, to certain living fresh-water siliceous protozoans ("Kiesel-Infusorien"). See Christian Ehrenberg, "Über das Massenverhältnis der jetz lebenden Kiesel-Infusorien und über ein neues Infusorien-Conglomerat als Polirschiefer von Jastraba in Ungarn," *Abhandlungen der Königlichen Akademie der Wissenschaften zu Berlin* (1836): 109–36. Since radiolarians are marine animals, probably Ehrenberg had observed living species of the class *Heliozoa*, which also have a silica skeleton but are fresh-water. In 1847, Ehrenberg described fossilized silica conglomerates from Barbados. He called them "Polycystinen" and identified, on the basis of their skeletons, 282 species, arranged in 44 genera. See Christian Gottfried Ehrenberg, "Über die mikroskopischen kieselschaligen Polycystinen als mächtige Gebirgsmasse von Barbados," *Monatsbericht der Königlichen Akademie der Wissenschaften zu*

Kölliker to Nice, they had unexpectedly met Johannes Müller, who had been collecting there. At that time, Müller had been working on the radiolaria, and he himself returned to St. Tropez in 1857 to complete his research. Müller's short monograph on these animals was his final publication, appearing just after his death.[41] Haeckel had the foresight – or perhaps just a simple desire for remembrance – to bring the tract with him to Italy. During the course of his own research, the monograph became his "gospel," and he virtually memorized it.[42] But Müller's work, it was clear, had been preliminary, and much remained for an ambitious researcher to do – especially to provide concrete meaning for that ever-nebulous claim of systematists that the several groups of organisms they treated were more closely or distantly *related*. When Haeckel produced his own monograph on the radiolaria – greater in

Berlin (1847): 40–60. Thomas Henry Huxley, while serving on board H. M. S. *Rattlesnake*, discovered what he thought to be a hitherto unknown zoophyte, which he called Thalassicolla (i.e., sea-jelly). Huxley skimmed connected masses of these one-celled creatures from the surface of the ocean. He noticed that glassy spiculae would sometimes be found along the surface of a cell. See Thomas Henry Huxley, "Zoological Notes and Observations Made on Board H. M. S. Rattlesnake during the Years 1846–50" (1851), in M. Foster and E. Lankester, eds., *The Scientific Memoirs of Thomas Henry Huxley*, 4 vols. (London: Macmillan, 1898), vol. 1, pp. 86–95. Huxley probably observed two related orders of the class of radiolaria now called Spumellaria – the Colloidea and the Beloidea. (Thalassicollida being a family of Colloidea). These orders either lack or have imperfect skeletons. Johannes Müller built upon the observations of Ehrenberg and Huxley in papers he read before the Berlin Academy of Sciences in 1855. In those papers, he confirmed Huxley's observations of the Thalassicolla, and because of the associated spiculae suggested that they might be related to sponges, on the one hand (which also have silica spiculae), and, on the other, to Ehernberg's Polycystina – Müller had found living specimens of these in waters off Messina in 1853. See Müller's "Über Sphaerozoum und Thalassicolla" and "Über die im Hafen von Messina beobachteten Polycystinen," in *Bericht über die Verhandlungen der Königlichen Preussischen Akademie der Wissenschaften zu Berlin* (1855): 229–54, 671–6. See also the following note.
[41] Johannes Müller, "Über die Thalassicollen, Polycystinen und Acanthometren des Mittelmeeres," *Abhandlungen der Königlichen Akademie der Wissenschaften zu Berlin* (1858): 1–62. As with Huxley, Müller described the Thalassicollia as without skeleton, or with skeleton only imperfectly represented (see the previous note). The Polycystina, which Ehrenberg had identified in 1847, displayed the silica skeleton, and the Acantharia, which are now usually distinguished as a related class (both under the subphylum Sarcodina), also had a skeleton, but not of silica. Müller called them all by the common name "Rhizopoda radiaria" or "radiolaria," and regarded them as closely related to other Rhizopoda, such as the amoeba – a common judgment made today. Müller divided the radiolaria into two major groups, those living singly and those colonially. The Thalassicolla, Polycystina, and Acantharia lived separately; but the first two also had colonial forms, called respectively Sphaerozoum and Collosphaera. For the most detailed modern study of these creatures, see O. Roger Anderson, *Radioloaria* (New York: Springer, 1983).
[42] Ernst Haeckel to Anna Sethe, February 29, 1860, in Haeckel, *Briefe an die Braut*, p. 163.

length and breadth of consideration, more beautiful by far than that of his teacher – he dedicated it to Müller, so that natural piety linked Müller's tragic end with Haeckel's glorious beginning.

Haeckel wrote to Anna to describe the creatures that would become his constant companions, though at one-thousandth to eight-hundredths of an inch in diameter they were hardly companionable:

The radiolaria are almost exclusively pelagic animals, that is, they only live swimming on the surface of the deep sea. . . . Their body consists of a hard and a soft part. The hard part is a siliceous skeleton, the soft is mostly a spherical, small, round capsule surrounded on all sides by an outcrop of many hundreds of exceptionally fine filaments, by which the animal moves and nourishes itself.[43]

Under his microscope ever new radiolarian species began to appear, so that by the spring he was able to ship back to Berlin specimens of some 101 species never before described.[44] (From the material dredged up during the expedition of the *Challenger*, which traveled around the world in the 1870s, Haeckel added several thousand more radiolarian species to his catalogue.)

Shortly after returning to Berlin, at the end of April 1860, Haeckel arranged to work on his collection at the Berlin Zoological Museum, where he had earlier cultivated a circle of friends and patrons, including the director, Wilhelm Peters (1815–1883), and the eminent Christian Ehrenberg (1795–1876), presiding secretary of the Berlin Academy of Sciences. Initially Haeckel prepared a report on his radiolarian work that Peters presented to the Academy of Sciences.[45] The report carefully described the new species he had discovered and analyzed their internal structure, something never before done. Haeckel's work remains today the starting point for further explorations with the scanning electron microscope. He determined that radiolarians had a soft body consisting of a central capsule, with a minute inner vesicle (*Binnenblase*), and surrounded by smaller vesicles (*Bläschen*), through which radiated a great number of stiff, threadlike pseudopodia. Depending on the family, the skeleton either surrounded the central capsule (as with the solitary Polycystinae or Eucyrtidium, Figure 6.2) or extended into the capsule (as with the Acanthometra and the colonial Polycystinae).[46] All

[43] Ernst Haeckel to Anna Sethe, February 29, 1860, ibid., pp. 161–2.
[44] Ernst Haeckel to Hermann Allmers, May 14, 1860, in Koop, ed., *Haeckel und Allmers*, p. 45.
[45] Ernst Haeckel, "Über neue, lebende Radiolarien des Mittlmeeres," *Monatsberichte der Königlichen Preussische Akademie der Wissenschaften zu Berlin* (1860): 794–817, 835–45.
[46] Ibid., pp. 795–7.

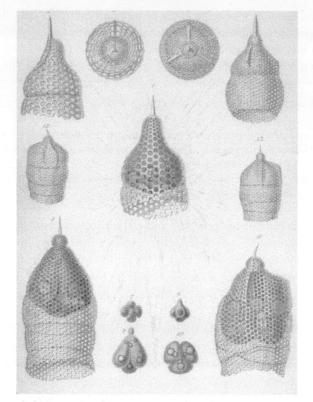

Figure 6.2. Radiolarian species of the genus *Eucyrtidium*, from Haeckel's *Radiolarien*, 1862.

of this was reiterated, with an elaboration of the systematics of the known species, in Haeckel's *Habilitationschrift*, rendered into Latin and completed in 1861.[47]

Yet neither the readers of the Academy report nor of the *Habilitationschrift* would have been prepared for the large two-volume monograph that Haeckel produced in 1862, his *Die Radiolarien (Rhizopoda Radiaria)*. The first two exercises announced a scholar of competence and promise, the latter showed the promise already brilliantly fulfilled. The monograph, which so astonished Darwin when he received it from Haeckel and which would be awarded the prestigious gold Cothenius medal of the Leopold-Caroline Academy of German Scientists (1863), displayed through the over 570 pages of the first

[47] Ernst Haeckel, *De Rhizopodum finibus et ordinibus* (Berlin: Reimer, 1861). The dissertation reappeared essentially as part IV of his large monograph on the radiolarians. See Ernst Haeckel, *Die Radiolarien*, 2 vols. (Berlin: Reimer, 1862), vol, 1, pp. 194–212.

volume and the 35 copper plates of the second many extraordinary features. I will mention just a few of the more significant.

First, with his discoveries Haeckel increased by almost half the number of known species of radiolarians. Second, he provided the most careful description of the distinguishing characteristics of the skeletons and soft parts, including extraordinarily exact measurements, given his instruments. His discrimination of the central capsule and the associated smaller vesicles, as mentioned earlier, set the foundation for later anatomical research.[48] Third, in anticipation of the kind of chorological considerations he would develop in later work, he specified the various seas in which a given species lived and the depths at which it could be found.[49] Fourth, and of considerable significance, he attempted to arrange his species into a "natural system" based on homology.[50] The two principal comparative axes for homological arrangement concerned the relation of the skeleton to the central capsule (either completely external to it, or partly inside it) and the forms of the skeleton itself (or its absence). On this basis, Haeckel distinguished, as they fell into pattern, some fifteen natural families.

Haeckel said he was inspired to attempt a natural system by the extraordinary book he had read while preparing his specimens – Charles Darwin's *Origin of Species* (1859). Haeckel first looked into Bronn's German translation of Darwin while at the Berlin Museum in the summer of 1860, just after he had returned from Messina. Being an anti-authoritarian – in his later days, to the point of dogmatism – Haeckel was probably enticed to read the new work because Ehrenberg and Peters both regarded it as a "completely mad book."[51] Though anti-authoritarian, Haeckel was not foolish; so it is not surprising that no mention of Darwin appeared in his Academy report in the fall or in his *Habilitationschrift*. It may be that the full impact of the *Origin* had not struck home during the composition of those pieces. In November 1861, while laboring full bore on his monograph, he again opened up the *Origin* and, as he related to Anna, "buried" himself in it.[52] From that fertile ground he emerged newly born for Darwin's theory, and the zeal of his conviction never cooled through later years.

[48] Haeckel, *Radiolarien*, vol. 1, pp. 68–116.   [49] Ibid., pp. 166–93.
[50] Ibid., pp. 213–40.
[51] Else Jahn, "Ernst Haeckel und die Berliner Zoologen," *Acta Historica Leopoldina* 16 (1985): 75.
[52] Ernst Haeckel to Anna Sethe, November 4, 1861, in Haeckel, *Himmelhoch Jauchzend*, p. 250. Haeckel's copy of Bronn's translation of Darwin's *Origin of Species* bears reading marks throughout. The copy is kept at Haeckel-Haus, Jena.

What kept Haeckel's enthusiasm for evolutionary theory glowing was the special contribution he thought he could make to establishing it empirically. He seems to have been especially piqued in this respect by Darwin's translator, the great paleontologist and morphologist Heinrich Georg Bronn (1800–1862). Bronn had added a concluding chapter to his translation of the *Origin* in which he evaluated the merits of Darwin's accomplishment. He had high praise for the ingenuity and provocative character of the hypothesis. Yet, he declared, it remained just that – a hypothesis, only a possible scenario of life's history:

We have therefore neither a positive demonstration of descent nor – from the fact that [after hundreds of generations] a variety can no longer be connected with its ancestral form (*Stamm-Form*) – do we have a negative demonstration that this species did not arise from that one. What might be the possibility of unlimited change is now and for a long time will remain an undemonstrated, and indeed, an uncontradicted hypothesis.[53]

Haeckel believed that he could provide the required positive proof of descent. Through the next decade and a half he would cultivate evidence of great power that he thought would strengthen Darwin's original conception, as well as lead to further important theoretical articulations. His appetite for this endeavor, though, was first sharpened by his radiolarian work.

In *Die Radiolarien*, Haeckel boldly sided with the English scientist. He argued that the radiolaria provided the desired empirical support for the new theory of evolution, since the relatedness of species within families bespoke genealogy and the transitional species joining families seemed to confirm it.[54] He even suggested that one genus, the *Heliosphaera*, might be regarded as the ur-type (Figure 6.3), since its symmetrical morphology and fundamental structure suggested how it might have been transformed into the other types.[55] In this light, Haeckel constructed a genealogical table that indicated the kind of descent relations these animals might actually express.[56]

Haeckel's adoption of Darwin's theory was made smooth by reason of three features of his intellectual situation. First, the actual fact of several intermediate species forms between the major groups of radiolaria begged for an evolutionary interpretation. Second, Haeckel's still-revered teacher, Rudolf

---

[53] H. G. Bronn, "Schlusswort des Übersetzers," in Charles Darwin, *Über die Entstehung der Arten im Thier- und Pflanzen-Reich durch natürliche Züchtung, oder Erhaltung der vervollkommneten Rassen in Kampfe um's Dasyn*, 2nd ed. (based on 3rd English ed.), trans. H. G. Bronn (Stuttgart: Schweizerbart'sche Verhandlung und Druckerei, 1863), p. 533.
[54] Haeckel, *Radiolarien*, vol. 1, pp. 231–3.   [55] Ibid., p. 233.
[56] Ibid., p. 234.

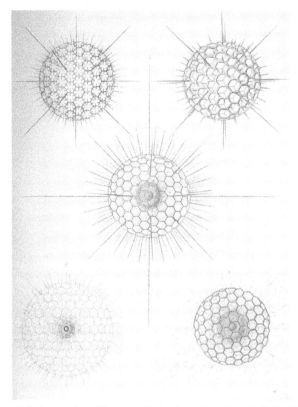

Figure 6.3. Radiolarian species of the genus *Heliosphaera*, from Haeckel's *Radiolarien*, 1862.

Virchow, had in 1858 declared that the mechanistic view of life, which he believed the only scientific outlook, required the conception of a transmutation of species.[57] Finally, the morphological tradition in which Haeckel was schooled, with its emphasis on homology, could easily be turned to

---

[57] Virchow had advanced the transmutation hypothesis tentatively in a lecture at the thirty-fourth meeting of German Naturalists and Physicians at Karlsruhe in September 1858. The lecture, entitled "Über die mechanische Auffassung des Lebens," was reprinted in Rudolf Virchow, *Vier Reden über Leben und Kranksein* (Berlin: Reimer, 1862). The lecture in this latter printing sparked the ire of M. J. Schleiden. The relevant passage concerning transmutation reads (p. 31):
Our experience justifies us in not holding as an inviolable rule good for all time that species are unchangeable, which at present seems so certainly to be the case. For geology teaches us to recognize a certain progression in which one species follows upon another, the higher succeeding the lower; and though the experience of our time opposes this, I must recognize that it seems to me a requirement of science that we return again to the transmutability of species. The mechanistic theory of life will thus obtain real security by taking this path.

evolutionary account. That morphological tradition also makes comprehensible why Haeckel would choose the genus *Heliosphaera* as the type of the
progenitor of the phylum. Most morphologists – such as Bronn and Haeckel's
friend Victor Carus, both of whom translated the *Origin* – emphasized that
the most symmetrical animal form (within particular constraints) served as
the fundamental type whence the other forms could be conceptually derived
by regular transformations.[58] The quite spherical *Heliosphaera actinota* thus
seemed to Haeckel the probable ur-species of the phylum (Figure 6.3).

Throughout his career, Virchow felt ready to endorse transmutationism as a viable and most
probable hypothesis; he balked, however, when zoologists, such as his one-time student
Haeckel, took it as a demonstrable fact.

[58] See, for example, J. Victor Carus, *System der Thierischen Morphologie* (Leipzig: Wilhelm
Engelmann, 1853); and H. G. Bronn, *Morphologische Studien über die Gestaltungs-Gesetze der
Naturkörper überhaupt und der Organischen insbesondere* (Leipzig: Winter'sche Verlagshandlung, 1858). Carus argued (p. 484), as most usually did, that these ur-types were only ideals
employed to represent actual animal forms; the actual forms, however, would always deviate
from the ideal. After reading Darwin, however, Carus made the ideal the real: a group exemplified a common form because it stemmed originally from a common ancestor bearing the
form. Bronn never quite came around to Darwin's particular idea that natural selection was
the unique power that produced new species from old. Prior to the publication of the *Origin*,
Bronn issued two books that set out a theory of progressive development of animal and plant
forms, though one that denied genealogical relationships of species and that postulated an
unknown natural law that ultimately expressed the plan of the Creator. Bronn's own extensive
paleontological studies persuaded him, contrary to the views of Lyell, that fossils bespoke
a progressive replacement of less complex creatures with more complex ones; such replacement, he contended, occurred in relation to the ever-increasing heterogeneity of geological
and climatological environments. This meant that a progressive complication of the physical
world stood as the harbinger of a progressive development of the biological, with each creature
fitting into the environment for which it had been adapted. Bronn thought it unscientific and
against all analogy with the physical sciences, however, to attempt to explain progressive introduction of new species through the direct actions of the Creator. "We rather believe that all
plant and animal species have been originally formed through an unknown natural force; but
they have not arisen through the reconstruction [*Umbildung*] of a small number of primitive
forms [*Urforme*]. That force stands in the closest and most necessary connection to those
forces and events forming the surface of the earth." See H. G. Bronn, *Untersuchungen über die
Entwichelungs-Gesetze der organischen Welt* (Stuttgart: Schweizerbart'sche Verlagshandlung,
1858), p. 82. Darwin's *Origin of Species* would seem to have offered Bronn just that "unknown
natural force" for which he was searching. However, in the "Schlusswort des Übersetzers" of
his translation of the *Origin*, he reiterated his former conviction. Additionally, he enumerated
several difficulties that made Darwin's hypothesis, granted as "possible," nontheless unlikely.
For instance, Darwin postulated multiple and very small changes in a variety as the initial
stage of a new form. But the many small features, all of which must contribute to an integrated
form, would change independently and at random, producing not a coherent type but only
a confusion having no advantage over competitors (pp. 534–44). Bronn cultivated these and
other difficulties that would later become canonized as among the principal objections to
Darwin's theory.

Aside from the evidence of family relations among the radiolaria – and the insight provided by reading Darwin – other more subjective, personal reasons may have inclined him to cast his lot with the new theory. In a long footnote to the section that showed how his work supported Darwin's, he referred to a clarion passage at the conclusion of the *Origin* in which the English scientist had issued a call to all the up-and-coming young naturalists to judge his ideas without prejudice. The note indicates that one zealous young iconoclast heard the resounding message:

I cannot let this opportunity pass without giving expression to the considerable aston-ishment I felt over Darwin's exciting theory about the origin of species. I am moved to do this even more because of the German professionals have found this epoch making work to be an unhappy presumption; they make this charge partly because they seem to misunderstand his theory completely. Darwin himself wished that his theory might be tested from every possible side and he looked "with confidence toward the young and striving naturalists who will be able to judge both sides of the question without partiality. Whoever is inclined to view species as changeable will, through the consci-entious admission of his conviction, do a good service to science; only thereby can the mountain of prejudice under which this subject is buried be generally avoided." I share this view completely and believe for this reason that I must express my convic-tion that species are changeable and that organisms are really related genealogically. Though I have some reservations about extending Darwin's insight and hypothesis in every direction and about all his attempts to demonstrate his theory, yet I must admire in his work the first, earnest and scientific effort to explain all appearances of organic nature from one excellent, unitary view point and his attempt to bring all sorts of inconceivable wonders under a conceivable law of nature. Perhaps there is in Darwin's theory, as the first effort of this sort, more error than truth. . . . The greatest confusion of the Darwinian theory lies probably herein, that it does not rest upon the origin of the urorganism – most probably a simple cell – whence all others have been developed. When Darwin assumes for this first species a special creative act, it seems of little consequence, and it seems to me not seriously meant. Aside from this and other confusions, Darwin's theory already has performed the immortal service of having brought the entire doctrine of relationships of organisms to sense and under-standing. When one considers how every great reform, every strong advance has found a mighty opposition, the more he will oppose without caution the rooted prejudice and battle against the ruling dogma; so one will, indeed, not wonder that Darwin's ingenious theory has, instead of well deserved recognition and test, found only attack and rebuff.[59]

Haeckel's support for Darwin's theory and his desire thereby to be ac-counted among the Darwinians would be reciprocated in Darwin's own dec-laration, some years later, that most of his ideas about human evolution (in *The Descent of Man*) had been antecedently confirmed by Haeckel – so in this, and

[59] Haeckel, *Radiolarien*, vol. 1, pp. 231–2, note 1.

other respects, Darwin could be accounted an Haeckelian.[60] But in Haeckel's radiolarian work, the feature that initially captured Darwin's admiration – and must that of any contemporary reader – was the artistic representation of the radiolarians.

Haeckel himself drew the figures that were transferred to the thirty-five copper plates used for printing the second volume of his book. His work required an extremely precise technique with the microscope, so small and delicate were these creatures. In order to get their intricate geometry correct, he would stud a potato with small rods, and then stabilize his model with the artist's sense of balance and proportion.[61] The principles of systematic display were not, however, genealogical, as perhaps might be expected from a budding evolutionist – thus the species *Heliosphaera actinota* (Figure 6.3), which served as the model of the Ur-type, came only with plate 9, not plate 1. The chief principle of ordering was, as he termed it, "natural"; but the order displayed a Goethean kind of nature, namely, the morphology of skeletalization. So, for instance, he began plate 1 with species of the genius *Thalassicolla pelagica*, which has no skeleton; plate 2 displays two aspects of *Aulacantha scolymantha*, which has some spiculae, and *Thalassicolla zanclea* and *Thalassolampe margarodes*, both of which lack any hard parts; plate 4 illustrates again *Aulacantha scolymantha*, along with forms (from different families and under-families) that have surrounding skeletons – namely *Prismatium tripleurum, Litharachnium tentorium,* and *Eucyrtidium lagena.* Yet, as is evident from these examples, principles other than the simply morphological governed his arrangement of the sequence of illustrations. Two further principles operated. The first was that of discovery – with a few exceptions, he represented only those species that he himself had found during his stays in Italy and Sicily; the already known species were simply described in the text. The second principle according to which he arranged his illustrations was aesthetic – each plate displayed conspicuous symmetries of form and color, and striking variability and individuality of these same qualities (see plates 2 and 4, Figure 6.4). This second principle prevented an organization that strictly followed the gradual

[60] In the Preface to his *Descent of Man and Selection in Relation to Sex* (2 vols. [London: Murray, 1871], vol. 1, p. 4), Darwin said of Haeckel's *Natürliche Schöpfungsgeschichte* (1868) that had it "appeared before my essay had been written, I should probably never have completed it. Almost all the conclusions at which I have arrived I find confirmed by this naturalist, whose knowledge on many points is much fuller than mine." Despite this avowal, to call Darwin a Haeckelian, as I have, will seem outrageous to some historians. I've tried to substantiate the sense in which this might be true in my *Meaning of Evolution: The Morphological Construction and Ideological Reconstruction of Darwin's Theory* (Chicago: University of Chicago Press, 1992).
[61] Ernst Haeckel to Anna Sethe, August 14, 1860, in Schmidt, ed., *Himmelhoc Jauchzend*, p. 133.

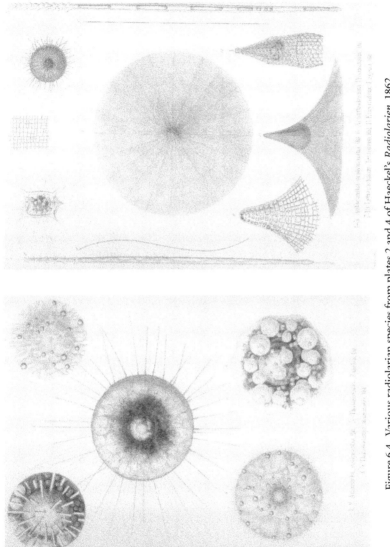

Figure 6.4. Various radiolarian species from plates 2 and 4 of Haeckel's *Radiolarien*, 1862.

gradual skeletalization of the animals. In later monographs, Haeckel's illustrations would more closely unite the morphological and the genealogical orders into one evolutionary tableaux of systematic arrangement. The considerations of form, color, and variability would, however, remain constant. The dramatic and exotic beauty of Haeckel's illustrations and their artful arrangement would in future play decided roles in persuading his readers of the evolutionary theory that would stand ever more strongly behind them.

CONCLUSION

Haeckel's habilitation had been a great success. And shortly after he published his two-volume radiolarian book in March of 1862, he was advanced to extraordinary professor at Jena. With his professional and financial security established, he was able to marry Anna, the individual whose love had made his success possible. Immediately he began lecturing on Darwin's theory, both at Jena and at various professional meetings. He dared to send a copy of his book to Darwin. He followed his gift with a letter that sketched the steps of his conversion to the new theory and indicated its spread in Germany. The letter was sent July 9, 1864, and provides a vivid indication of the fidelity of his new faith.

My dear Sir,

I found your letter [thanking him for the gift of his book], which had been written several months ago, when I returned from a zoological trip to the Mediterranean. Your letter has given me great pleasure. It has also provided me opportunity and personally the decided honor, Sir, to express the extraordinary esteem I have for the discoverer of the "Struggle for Life" and "Natural Selection." Of all the books that I have read, none has made so powerful and marked an impression on me as your theory of the origin of species [*Ihre Theorie über die Entstehung der Arten*]. In this book I find at once the harmonious solution to all the fundamental problems of which I have labored for an explanation since the time I had learned to know nature in her authentic state. Since then I have studied your theory – I say without exaggeration – daily, and whether I study the life of man, animals or plants, I find in your descent theory the satisfactory answer to all my questions.

Since you must have a certain interest to learn of the spread of your theory in Germany, allow me to impart this. Most of the older zoologists, and among them many of considerable authority, are among your most enthusiastic opponents. On the one hand, these men have lost, through a life spent in the old accustomed dogmas, the ability to view impartially what is new as worthy and correct – even truth itself; on the other hand, they lack the courage to admit their actual belief in the truth of the descent theory. Many attempt to correct their earlier false views and so, finally, are not able to comprehend the whole of nature with one overview, since the painstaking study of details and the analytic investigation of particulars does not permit a general perception of nature.

Yet among the younger naturalists, the number of your committed and enthusiastic followers grows from day to day; and I believe that in a few years their number will be as large, perhaps, as the number of your committed followers in England itself. The Germans on the whole (as far as I can judge) are not so constrained by religious and social prejudices as the English – though in respect of political maturity and in relation to full development they are rather behind. . . . The academic lectures, which I myself and a few of my younger colleagues conduct on your theory, appeal not only to students of natural science and medicine, but also are heard by philosophers and historians, and yes, even theologians. For the historians, a new world is opened, since in the application of descent theory to human beings (as Huxley and Vogt have so happily attempted), they find a way to connect closely the history of human beings with natural history. Indeed, it is here in Jena that we have particularly favorable ground for the development and spread of such reformational teaching, since in all respects we have here the greatest freedom – while at other universities – for instance Göttingen and Berlin – many restrictions and general rules hinder more free intellectual action. Yet one may hope that the progressive development that one hears has begun in all quarters of Germany will defeat, now and again, the opposing elements and that the results of your theory will be correctly understood and adopted.

Perhaps you will allow me to relate to you a few personal matters concerning your theory, since I have devoted my life to it and direct all my activities to making it known. I had decided to do so immediately after I came to know it. In my first major work, a monograph on the radiolarians (Berlin, Reimer, 1862), I mentioned your theory along the way (p. 232, in the note), and attempted to construct a genealogical table of the relationships of these animals. Then, last year, I seized the opportunity in Stettin for the first time, at the meeting of the German Naturalists, to bring the question into the discussion; this resulted in a rather lively debate. Though I was strongly attacked by a very eloquent speaker, Dr. Otto Volger from Frankfurt, I yet won many friends for your theory and Virchow, our greatest scientific physician, also spoke favorably on it.

Presently I am busy with a large work on coelenterates, the animals which, because of their complicated sort of development, show very well their common descent from one original form. On the coast of Nice this spring, I spent a long time studying Medusae. I was astonished at the extraordinary spread of individual variations that occurs in some of these animals. One often finds the formation of the essential parts of individuals of one and the same species to be greater than those between different species of a genus and, indeed, between several genera of one and the same family. With your permission, I will send you next year my work on this subject. . . .

Although I am only 30 years old, a terrible fate, which has destroyed my whole happiness in life, has made me mature and resolute. It has hardened me against the blame as well as the praise of men, so that I am completely untouched by external influence of any sort, and only have one goal in life, namely to work for your decent theory to support it and perfect it.

Please forgive me, Sir, for having taken up your precious time with this long letter. It was for me a vital necessity to express to you at least once those things that move me daily in my tasks and that suffuse all my work. "When the heart is full, the mouth overflows."

My friends and colleagues here, the comparative linguist August Schleicher and the comparative anatomist Carl Gegenbaur, with whom I so often share my strong conviction of the pure truth of your teaching – they send their best wishes. I hope, Sir, that your health improves and that for a long time you will be ready to fight the good fight for the truth and against human prejudice. I remain with the great respect, yours very truly,

Ernst Haeckel.[62]

In the conclusion of this letter, Haeckel hinted at a tragedy that had recently befallen him. On February 16, 1864, his thirtieth birthday, the same day that he received notice that he had been awarded the Cothenius medal from the Academy of German Natural Scientists for his radiolarian book, his beloved wife of eighteen months suddenly died of a mysterious fever. Haeckel was devastated, so much so that his parents and brother thought he might commit suicide. This tragedy would scar him for the rest of his life. Even in his late sixties, as his birthday would come due, he would experience that constantly renewed depression and again turn his thoughts to suicide. This tragedy also led him to replace a now-extinguished religious faith with a new kind of fervent conviction – Darwin's theory of evolution. In future, Haeckel not only would defend Darwinism against the infidels, he would pour hot vitriol on all who attacked the theory. Moreover, he would distance himself from those heretics who virtually eliminated the role of natural selection – such as Spencer – or who thought natural selection the only mechanism of species change – such as Weismann.

Contemporary scholars who yet perceive in Haeckel the apostate do so, I suspect, because of the way he absorbed Darwinian theory into his decidedly Romantic and volatile character. There is little doubt that Haeckel's personal crisis caused him to advance evolutionary theory with a determination and excess that stood in marked contrast to the more sober demeanor of the mature Darwin. But the intellectual components of his theory hardly differed from that of the master himself. The zeal, though, was distinctively his own.[63]

---

[62] Ernst Haeckel to Charles Darwin, July 7, 1864, held in the Darwin Papers, the Manuscript Room, Cambridge University Library.

[63] This essay is adapted from my *The Tragic Sense of Life: Ernst Haeckel and the Battle over Evolutionary Theory in Germany* (forthcoming).

CHAPTER SEVEN

Adaptive Landscapes and Dynamic Equilibrium

The Spencerian Contribution to Twentieth-Century American Evolutionary Biology

Michael Ruse

> It would be quite justifiable to ignore Spencer totally in a history of biological ideas because his positive contributions were nil.
>
> Ernst Mayr, *The Growth of Biological Thought*[1]

The standard history of evolutionary biology – the history I myself was writing some two or more decades ago[2] – runs something like this: In the *Origin of Species*, published in 1859, Charles Darwin tried to do two things. First, he wanted to establish the fact of evolution. Second, he proposed a mechanism for evolution – natural selection brought on by a struggle for existence. Darwin was successful in his first aim. Very soon after the publication of his book, most of the educated world – scientists and laypeople – was converted to evolutionism. It became the accepted way of thinking about life's origins, including our own. Darwin was unsuccessful in his second aim. Almost no one took up natural selection as a working cause of evolutionary change – rather, a host of alternatives were preferred, including Lamarckism (the inheritance of acquired characteristics), saltationism (evolution by jumps), orthogenesis (lines of development that take on their own momentum), and others. The triumph of selection as a mechanism had to wait until the twentieth century. It was only then that biologists made the required major advances in our

---

[1] Ernst Mayr, *The Growth of Biological Thought* (Cambridge, MA: Harvard University Press, 1982), p. 386

[2] For instance, in my *The Darwinian Revolution: Science Red in Tooth and Claw* (Chicago: University of Chicago Press, 1979).

understanding of the mechanisms of heredity and developed the science now
known as "genetics." In the tradition of Gregor Mendel, it was realized that
transmission is particulate, in the sense that the units of heredity get transmitted from generation to generation in virtually an unchanged form. Hence,
there is a basis of stability on which forces for change can operate. Building
on these new understandings about heredity, mathematically talented theoreticians generalized across groups and saw how natural selection can be, and
indeed is, the major force behind long-term evolutionary change. In particular, three people – three "population geneticists" – were responsible for this
advance: Ronald Fisher and J. B. S. Haldane in Britain, and Sewall Wright in
America. Finally, the naturalists and experimenters put empirical flesh on the
theoretical skeletons, and so the "synthetic theory of evolution" (a synthesis
of Darwin and Mendel), or "neo-Darwinism," was born. The second aim of
Charles Darwin was finally realized. Natural selection, the mechanism of the
Origin, was seen as the key to evolutionary change.[3]

Although a little rough-and-ready, this sketch still seems to me to be a basically accurate account of the history of evolutionism in Britain. One should
not neglect the fact that there were some immediate triumphs of selective
explanation, notably Henry Walter Bates's analysis of butterfly mimicry and
(somewhat later) Raphael Weldon's analysis of crab carapace dimensions. But
these were really exceptions rather than the rule. For real causal movement
forward, the key work after Darwin was Fisher's *The Genetical Theory of Natural Selection*, published in 1930. It is true that this book had some highly
nonscientific ideas lurking beneath the surface: Fisher's ardent eugenicism for
a start and his commitment to Anglican Christianity for a second. Nevertheless, it was a work – and was seen to be and appreciated as a work – that
made natural selection *the* evolutionary mechanism first and foremost.[4] As
an account of the history of evolutionism in America, however, I argue that
the sketch I've just given is totally misleading. I do not want to say that, in the
New World, Darwin's ideas had no influence in the first half of the twentieth
century and were less than fully realized in the second half – there are today
many first-class, fully committed American Darwinian evolutionists (meaning evolutionists for whom natural selection is the only mechanism that really
counts) – but I would say that, in the American version of this revitalized

[3] A classic formulation is given in Peter Bowler, *Evolution: The History of an Idea* (Berkeley:
University of California Press, 1982).
[4] Michael Ruse, *Monad to Man: The Concept of Progress in Evolutionary Biology* (Cambridge,
MA: Harvard University Press., 1996). Much of the background information for this chapter
can be found in my book.

evolutionism (Mendelized evolution, that is), Charles Darwin was not the major influence. I would go further and say that there was a major influence, acknowledged or not, and that this was Darwin's fellow English evolutionist Herbert Spencer.

I would argue indeed that, notwithstanding the concessions just made, we still find significant traces of Spencerian thought in American evolutionary biology today. But in this chapter I shall not attempt to make the overall case but rather restrict myself to one task only: showing the importance of Spencer's thinking on the evolutionary theorizing of the American contributor to the foundation of population genetics, Sewall Wright. This man, born in 1889 (and who died at the great age of ninety-eight), was trained at Harvard by the pioneering geneticist W. E. Castle, then worked for ten years at the United States Department of Agriculture, and finally went to the biology department at the University of Chicago, where in the early 1930s he published what he called the "Shifting Balance Theory of Evolution." I certainly agree with the conventional history of evolutionism that this theory played a key role in the subsequent development of evolutionary research in America, most importantly through its great influence on the thinking of the Russian-born evolutionist Theodosius Dobzhansky. The latter's *Genetics and the Origin of Species* (published in 1937) is rightfully acknowledged as the paradigm-creating work that led to almost everything that followed, including (paradoxically and amusingly) Ernst Mayr's wonderful work, *Systematics and the Origin of Species* (published in 1942). So, restricted though my aims may be, if I succeed in my task I will have done much to make the overall case at least plausible and worthy of further investigation.

COMPARING DARWIN AND SPENCER

Let us begin at the beginning. What was Charles Darwin's thinking on evolution? What was Herbert Spencer's thinking on evolution? In Darwin's case, it is best to begin with the central arguments for his mechanism of natural selection, given in early chapters of the *Origin*. First, Darwin argued for a struggle for existence:

A struggle for existence inevitably follows from the high rate at which all organic beings tend to increase. Every being, which during its natural lifetime produces several eggs or seeds, must suffer destruction during some period of its life, and during some season or occasional year, otherwise, on the principle of geometrical increase, its numbers would quickly become so inordinately great that no country could support the product. Hence, as more individuals are produced than can possibly survive, there must in every case be a struggle for existence, either one individual with another of the same species, or with the individuals of distinct species, or with the physical conditions of life. It is the doctrine of Malthus applied with manifold force to the whole animal and vegetable kingdoms; for in this case there can be no artificial increase of food, and no prudential restraint from marriage.[5]

Second, he moved on to argue for natural selection:

Let it be borne in mind in what an endless number of strange peculiarities our domestic productions, and, in a lesser degree, those under nature, vary; and how strong the hereditary tendency is. Under domestication, it may be truly said that the whole organization becomes in some degree plastic. Let it be borne in mind how infinitely complex and close-fitting are the mutual relations of all organic beings to each other and to their physical conditions of life. Can it, then, be thought improbable, seeing that variations useful to man have undoubtedly occurred, that other variations useful in some way to each being in the great and complex battle of life, should sometimes occur in the course of thousands of generations? If such do occur, can we doubt (remembering that many more individuals are born than can possibly survive) that individuals having any advantage, however slight, over others, would have the best chance of surviving and of procreating their kind? On the other hand, we may feel sure that any variation in the least degree injurious would be rigidly destroyed. This preservation of favourable variations and the rejection of injurious variations, I call Natural Selection.[6]

Backing the direct case for selection, Darwin also argued analogically, using the evidence and techniques of artificial selection, the work of the breeders of animals and plants for profit and pleasure – bigger and shaggier sheep, fleshier turnips, stronger bulldogs, and fancier pigeons. If we humans can do so much, he claimed, nature can do far better. And then, with the case made for selection, moving now to the arguments that really convinced people of the fact of evolution, Darwin trawled through the whole spectrum of biological studies, showing how his thinking throws light on so many different and diverse areas of interest and inquiry. Instinct, paleontology, biogeographical distribution, morphology, embryology, taxonomy, and more – all of these are made reasonable by evolution through selection, and conversely evolution through selection is justified and confirmed by its explanatory successes over such a wide area. As in the best theories – astronomy, optics, geology – there was at the heart of Darwin's thinking a "consilience of inductions," as it was termed by his former mentor, the philosopher William Whewell.[7]

With an eye to future discussion, a matter of some interest is where exactly Darwin stood on the subject of progress. Was Darwin committed to a

---

[5] Charles Darwin, *On the Origin of Species* (London: John Murray, 1859), p. 63.
[6] Darwin, *Origin*, pp. 80–1.
[7] William Whewell, *The Philosophy of the Inductive Sciences* (London: Parker, 1840).

view of evolution that saw an upward rise from the primitive and simple, the monad, to the sophisticated and complex, notably man? Although there is still some debate about this, the unequivocal answer is that Darwinism equivocates! Such an upward rise was always part of Darwin's own personal view of life's history. He did indeed believe in progress. However, with respect to his science – with respect to his evolutionism – progress was always somewhat problematical for Darwin. He did not see it as built into the process, and he could see that in some respects selection rather points in the other direction. There is a relativism to selection that is antithetical to progress. But Darwin wanted biological progress, and so he got it. He decided that progress does occur, and that the key functional phenomena are what today's evolutionists call "arms races" – lines compete against each other and eventually one wins, and even more eventually there is an absolute winner. In the language of today's most ardent Darwinian, Richard Dawkins, this winner is that organism with the biggest on-board computer, namely *Homo sapiens*.[8]

I am suggesting therefore that – for all his strong personal commitment to the idea – as a selectionist, Darwin felt that progress is added on, rather than built in. And this is a good point at which to introduce Herbert Spencer, for his evolutionism starts (continues and finishes) with progress. It is the very backbone of his thinking, to use an appropriate metaphor. For him, progress was not so much an empirical finding as a metaphysical presupposition of his view of history. It ran through everything, from the most primitive forms of culture to the evolution of our own species.

Now, we propose in the first place to show, that this law of organic progress is the law of all progress. Whether it be in the development of the Earth, in the development of Life upon its surface, in the development of Society, of Government, of Manufactures, of Commerce, of Language, Literature, Science, Art, this same evolution of the simple into the complex, through successive differentiations, holds throughout. From the earliest traceable cosmical changes down to the latest results of civilization, we shall find that the transformation of the homogeneous into the heterogeneous, is that in which Progress essentially consists.[9]

Note the great importance for Spencer of what we might call the organic metaphor. He thinks hierarchically – from cell, to organism, to state, to whole. He is quite explicit in thinking of the state as a kind of organism, and in thinking

---

[8] Richard Dawkins, *The Blind Watchmaker* (New York: Norton, 1986).

[9] Herbert Spencer, "Progress: Its Law and Cause," *Westminster Review* 67 (1857): 244–67; reprinted in *Essays: Scientific, Political and Speculative* (London: Williams and Norgate, 1868), vol. 1, pp. 1–60, at pp. 2–3.

that progress at one level is mirrored by progress at another level. Which brings up the question of causes or mechanisms. Here, Spencer showed an eclectic synthesis of German morphology and British thermodynamics, seasoned with a good dash of British nonconformist thinking on society and the desirable underlying economic forces, arguing (again perhaps more metaphysically than empirically) that nature starts in a condition of uniformity – what he called "homogeneity" – and tends naturally to a condition of complexity – what he called "heterogeneity." Why should this be so? Apparently it follows directly from the fact that causality tends to be open-ended, inasmuch as one cause leads to multiple effects, rather than many causes' leading to one effect. There is always a kind of explosion or expansion outward, as the simple and uniform tends toward the complex and diverse. This happens at all levels of the hierarchy – organisms, states, whatever. Something internal or external jogs or disturbs the state of being, and the multiplying causal process kicks in. More than this, however – for as the process of complexification is occurring, there is a tendency to move upward to a higher level of existence. Life – everything – is rather like the incoming tide, set on its end. There are surges forward, followed by moments or periods of consolidation, then further surges forward, with overall gain happening over and over again. Disturbance leads to the attempt to move back to a state of rest, but the new state is never the same as the old state – it is more heterogeneous, and higher. Overall, therefore, evolution can be described (as it came to be known) as an exemplification of "dynamic equilibrium."

Did Spencer have time for more mundane processes, such as natural selection? As it happens, after Darwin himself had discovered the process, although before Darwin moved into print, Spencer wrote of selection as a contributing factor to the evolutionary process.[10] But – although it was he who provided the alternative name of the "survival of the fittest" – selection was ever a secondary mechanism for Spencer. He opted for so-called Lamarckism as primary – that is, for the inheritance of acquired characteristics. Like Darwin, Spencer thought that the Malthusian explosion was important, but unlike Darwin, Spencer thought that the chief effect would be to spur organisms to greater effort, thus stimulating their evolution up the chain of being, as their simple forms transmuted into the more complex. Darwin himself, one might add, always had a place for Lamarckism, but for him it was selection first and Lamarckism second. For his fellow Englishman, it was Lamarckism first and selection second.

[10] Herbert Spencer, "A Theory of Population, Deduced from the General Law of Animal Fertility," *Westminster Review* 1 (1852): 468–501.

SOCIAL DARWINISM

Turn now to another important background issue. Darwin and Spencer alike always thought that their biological theorizing had implications for broader societal questions – questions about culture and about society and about gender relationships and about race and much more. One has only to read the *Descent of Man* to see how very important these issues were for Charles Darwin. One has only to read anything by Herbert Spencer to see how very important these issues were for Herbert Spencer. To be quite frank, there is more overlap in some of the biological-cum-social thinking of Darwin and Spencer than today's Darwinians are always happy to acknowledge. Darwin, for instance, has some very Victorian views on the virtues of the class structure and on capitalism and much more.[11] But although the move from biology to society became known as "social Darwinism," all historians properly agree that in many, if not most, respects it was Spencer who blazed the trail.[12] It was he who saw the processes of biology and culture, taken broadly, as being similar, if not identical. It was he who saw natural selection as being the biological equivalent of the societal process of laissez faire, and who urged socioeconomic nostrums in the name of the overall metaphysical processes of life. Not that Spencer necessarily always argued from biology to society – temporally and conceptually, it was often the other way. The point is that for him it was the big picture that counted. In what Spencer grandly called the "Synthetic Philosophy," it was evolution as a world picture that counted.

   And this last point was surely a major factor in Spencer's great general success in late Victorian Britain and (even more) in America in the postbellum years. Once people were over the initial shock of the idea of transmutation – and for many, this was a pretty minor shock – they became not only evolutionists but positive enthusiasts. Indeed, as I have argued at length elsewhere, one can properly say that evolution aspired less to being a science (just as well, for in this direction it was, at best, second-rate) and more to being something akin to a secular metaphysics or religion.[13] It was seen as the alternative to the Christianity that had failed (or for some, as the revitalizing force of the Christianity that had failed) and as something – with its story of origins and

---

[11] Michael Ruse, *The Darwinian Revolution: Science Red in Tooth and Claw* (Chicago: University of Chicago Press, 1979, 2nd ed. 1999).

[12] Robert J Richards, *Darwin and the Emergence of Evolutionary Theories of Mind and Behavior* (Chicago: University of Chicago Press, 1987).

[13] Michael Ruse, *Monad to Man*; Ruse, *Mystery of Mysteries: Is Evolution a Social Construction?* (Cambridge, MA: Harvard University Press, 1999); Ruse, *The Evolution Wars* (Denver: ABC-CLIO, 2000).

its drama of the human rise to the top – that answered all of the questions left hanging by the inadequate myths of the past. Good religions – that is to say, the monotheistic religions of the West – have social implications and promote ethical dictates about proper behavior. Love your neighbor as yourself, and so forth. Given Spencer's vigor in this direction, his positive embrace of an overall view of the evolutionary process, one that is ever-moving and ever-surging forward and one that pushes humans higher and yet higher, it is hardly therefore surprising that it was he – far more than Charles Darwin – who came to epitomize the evolutionary way of thinking.[14] It was his books to which one turned first; it was he who was taken as the definite authority on matters of morality and custom and proper behavior – in the private and the public spheres. Nor should one think that Spencer's influence was restricted only to one segment of society, specifically, to that segment that favored free economic competition – success to the winner, and widows and children to the wall. It is the first axiom of religion that true believers rarely, if ever, achieve uniformity of belief, especially in matters of morality – for every Quaker or Mennonite pacifist, there is a military chaplain urging one on to kill in the name of Jesus. Likewise with social Darwinism. Businessmen liked unrestrained competition and justified this liking in the name of evolution. Bureaucrats liked organization and state control, and they too justified themselves in the name of evolution. Even the Marxists got on the bandwagon, for their theorizing often owed far more to Spencer and his works than it did to Marx and *Capital*.[15]

Many were the American children who received a copy of one of Herbert Spencer's works on school prize day. Many were the young men who joined discussion groups to argue over the Synthetic Philosophy. Many were the rich men who bought their way into the Kingdom of Heaven through the promotion of heterogeneity over homogeneity. Many were the poor men who fought the bosses in the name of dynamic equilibrium.[16] This being so – as we leave the nineteenth century and turn toward and into the twentieth – there

---

[14] See, for instance, the fascinating discussion of Spencer's influence in the Far East in J. R. Pusey, *China and Charles Darwin* (Cambridge, MA: Harvard University Press, 1983).

[15] Mark Pittenger, in *American Socialists and Evolutionary Thought: 1870–1920* (Madison: University of Wisconsin Press, 1993), discusses in some detail the extent to which Spencer influenced the thinking of all levels of American society, not just the traditionally recognized businessmen.

[16] D. Duncan, *Life and Letters of Herbert Spencer* (London: Williams and Norgate, 1908), shows well the great influence of Spencer, something reinforced by secondary works such as C. E. Russett, *The Concept of Equilibrium in American Social Thought* (New Haven, CT: Yale University Press, 1966).

is one very obvious prediction. One would expect to see in American evolutionary biology the influence of Herbert Spencer. Earlier on, in the immediate post-*Origin* period, the greatest influence was probably a vehement anti-evolutionist – the Swiss transplant and Harvard professor Louis Agassiz. Educated by the *Naturphilosophen* (with a strong dash of Georges Cuvier), Agassiz could never accept species change, but all of his students (including his own son) went over the divide. And as they went, they took with them all sorts of beliefs about archetypes and homologies and upward change and more. But even from the first, Spencer started to fuse in, which was hardly that surprising, since it is probable that he himself drew on at least some German sources (never an easy matter to decide, given Spencer's reluctance to acknowledge the influence of others). Edward D. Cope, the great paleontologist, was open about the influence of Spencer.[17] And more and more the same started to be true of others, as the century progressed and as the Englishman's fame and writings spread, as what professional evolution there was became increasingly frustrated by inadequate techniques and theory, and as evolution's public status as a secular religion became yet more firmly established and acknowledged.[18]

Consider the position of E. G. Conklin – major cytologist, a great influence on the structure and running of American biology, and a frequent writer on evolution:

Life itself, as well as evolution, is a continual adjustment of internal to external conditions, a balance between constructive and destructive processes, a combination of differentiation and integration, of variation and inheritance, a compromise between the needs of the individual and those of the species. And in addition to these conflicting relations we find in man the opposition of instinct and intelligence, emotion and reason, selfishness and altruism, individual freedom and social obligation. Progress is the product of the harmonious correlation of organism and environment, specialization and co-operation, instinct and intelligence, liberty and duty.[19]

"Adjustment of internal to external"; "balance between constructive and destructive"; "combination of differentiation and integration"; "Progress is

---

[17] See H. F. Osborn, *Cope: Master Naturalist: The Life and Writings of Edward Drinker Cope* (Princeton, NJ: Princeton University Press, 1931).

[18] I document these general claims in my trilogy on science and values, *Monad to Man: The Concept of Progress in Evolutionary Biology* (Cambridge, MA: Harvard University Press, 1996); *Mystery of Mysteries: Is Evolution a Social Construction?* (Cambridge, MA: Harvard University Press, 1999); and *Darwin and Design: Science, Philosophy, Religion* (Cambridge, MA: Harvard University Press, 2003).

[19] E. G. Conklin, *The Direction of Human Evolution* (London: Oxford University Press, 1921), p. 87.

the product of the harmonious correlation..." If this is not pure Herbert Spencer, I don't know what is.

But you may interject that all of this is true of evolutionary biology only while it was still in its prescientific (or second-class scientific) state, before it matured, thanks to the efforts of the great theoretical population geneticists. Once Fisher, Haldane, and (especially in our case) Sewall Wright had set to work, evolution as a science was moved right up – and one of the major features of this move up is that Spencerianism must have been expelled. From then on (say, around 1930), it was pure Darwinism all the way. No one would deny Spencer at the beginning of the century, just as no one should deny Darwin before the middle of the century. Yet is this really true? I should say (by way of warning) that proving a positive case – that Spencer was indeed an important influence – cannot be an easy task. Along with the intellectual maturing, which came with the arrival of population genetics, came the social drive to professionalize evolutionary theory. No longer was evolutionism to be something functioning primarily as a secular substitute for Christianity, confined (as it then was) to museums and to the popular lecture hall and to the magazine for the general reader. It was to be part of real science, taught in universities, done by trained specialists, getting students and grants. But one of the chief marks of professional science is that it stays strictly away from religion and "philosophy," where something like social Darwinism –with its overt moral and political agenda – is precisely what is meant and feared. Hence, even if Spencer was an influence, from about 1930 on – given especially that Social Darwinism was now being tied into some of the worst excrescences of the twentieth century (just look at *Mein Kampf*, for starters) – one should expect a reluctance to parade the fact too obviously. Given that Ernst Mayr was one of those who worked longest and hardest to upgrade evolutionary biology, it is little wonder that the history he wrote was one that included the sentiment expressed at the beginning of this chapter.

THE SHIFTING BALANCE THEORY AND ITS DISCONTENTS

But enough of pouring water on the altar. Can it be lit? Even though the time when Mendel was coming into evolutionary biology was also just the time when evolutionists would be eager to deny or downplay the influence of Spencer, can we nevertheless establish the continued influence of Spencer? Specifically, can we establish the continued influence of Spencer on and in Sewall Wright's Shifting Balance Theory of evolution? To start the ball rolling, let us begin by noting a number of interesting (and I shall argue, significant) facts about this theory.

First, no one could follow the theory. True, this is a bit of an exaggeration – but not much. The theory is presented in two places: first, in a long and rather technical paper in *Genetics* in 1931; second, in a much shorter poster paper – a *Reader's Digest*–type condensation – given at an international congress on genetics in 1932.[20] It is the first paper that lays out the guts of the theory. It is the second paper that people read and thought they understood. No one could follow the math of the first paper. Dobzhansky openly confessed that Wright's calculations were beyond him.[21] Later when he and Wright coauthored papers, Dobzhansky admitted almost proudly that he understood the first and last lines only. And the same blankness toward theory was characteristic of the other important evolutionists of the period. Mayr has never been able to follow one symbol next to another. Again and again he declaims against formal techniques and products.[22] And others were little better. Some years ago, I interviewed the botanist of the group, G. Ledyard Stebbins, the author of *Variation and Evolution in Plants*. He saw the Wright paper at the congress and was at once very excited (or, if memory was improving on the occasion, certainly became very excited). In the same interview, Stebbins admitted without hesitation or attempt to conceal that the reason for the excitement was that here was something he could understand – no math.[23]

Second, the theory – especially the theory as given in the second paper – was seriously confused. It was really seriously confused, to the point of incoherence. The key notion of this second paper was that of an adaptive landscape. This metaphor enabled Wright to collapse down a huge amount of information into an easily graspable, visual picture. The picture was supposed to be in three dimensions, with x and y axes showing where one finds organisms (or are they groups?), and the z axis (generally not shown in a two-dimensional picture) sticking out and representing fitness. The higher up the landscape, the fitter the inhabitant. Supposedly, organisms sit on the tops of peaks, basically kept up there by selection. Every now and then, however, a population wanders off the top, down the side, and, if lucky (most are not), up the side of another mountain or hill. How does this happen? Through the random effects of breeding, where small-population contingencies outweigh selection – in other

[20] Sewall Wright, "Evolution in Mendelian populations," *Genetics* 16 (1931): 98–160; Wright, "The Roles of Mutation, Inbreeding, Crossbreeding and Selection in Evolution," *Proceedings of the Sixth International Congress of Genetics* 1 (1932): 356–66.

[21] E. Mayr and W. Provine, eds., *The Evolutionary Synthesis: Perspectives on the Unification of Biology* (Cambridge, MA: Harvard University Press, 1980).

[22] See E. Mayr, *Towards a New Philosophy of Biology: Observations of an Evolutionist* (Cambridge, MA: Harvard University Press, 1988).

[23] Interview with the author, May 25, 1988.

words, through so-called genetic drift. Again, supposedly, the reason why a population might suddenly start moving up, after a drift-driven downward journey, is that a new combination of features might get together, and this (or these) might prove adaptively advantageous. Finally, some organism or group that was much better than others on nearby peaks would either beat everyone else, or its fancy new features would get spread around. The process is over until the next time.

Fine and dandy, except – as Will Provine, Wright's dedicated and splendid biographer, showed – there is really radical confusion about those x and y axes.[24] What are we actually plotting? Is it gene frequencies? If so, which genes and how and why? Can one simply split everything apart in this kind of reductionistic fashion, treating the genes like (to use Ernst Mayr's metaphor) beans in a bag? Or is it all a matter of individual genotypes, so that points on the graph represent individual organisms? In which case, why can one assume a smooth transition from one genotype to another? As Provine pointed out, the trouble is that (in order to get his adaptive landscape) Wright is indeed collapsing a huge amount of information into two or three dimensions. The virtue is that a lot of information can now be presented very simply. The mis-virtue is that, not only is a lot of information lost, a lot of information is confused. Wright himself had apparently never even thought about these issues until he was alerted to them. But even he had to agree that, when you start to peer into what the adaptive landscape is all about, it is a conceptual mess.

Third, the theory is false from beginning to end. It has virtually no connection at all with the real world. Recent analyses of the theory – theoretical and empirical – show that it just does not work or do the job required of it – namely, to explain change, in an adaptive fashion or otherwise. Properly characterizing the view of Fisher as involving a selective force working on large populations, being driven to adaptive excellence, the most severe critics write as follows:

Although the mathematics of the shifting balance theory (henceforth SBT) is complicated, its essence is simple. Wright proposed that adaptation involved a shifting balance between evolutionary forces, resulting in a three-phase process:

Phase I: Genetic drift causes local populations (demes) to temporarily lose fitness, shifting across "adaptive valleys" toward new "adaptive peaks."

Phase II: Selection within demes places them atop these new peaks.

---

[24] William Provine, *Sewall Wright and Evolutionary Biology* (Chicago: University of Chicago Press, 1986).

Phase III: Different adaptive peaks compete with each other, causing fitter peaks to spread through the entire species. (Wright believed that populations occupying higher adaptive peaks would send out more migrants, ultimately driving other populations to the highest peak.)

There is thus a clear distinction between the Fisherian and Wrightian views of evolution: the former requires only that populations be larger than the reciprocal of the selective coefficient acting on a genotype, and the latter requires subdivided populations, particular forms of epistasis, genetic drift that counteracts selection, and differential migration between populations based on their genetic constitution.[25]

They write then:

We begin our analysis with an examination of the theory itself and then discuss the data offered in its support by Wright and others. We will conclude that (1) many of Wright's motivations for the SBT were based on the problems he perceived with the alternative process of mass selection, but these problems are largely illusory; (2) although, as Wright postulated, alternative adaptive peaks separated by adaptive valleys clearly exist, there is little evidence for the assumption that movement between peaks involves a temporary loss of fitness; (3) although phases I and II of the theory may be at least theoretically plausible, there is little theoretical support for phase III of the shifting balance, in which adaptations spread from particular populations to the entire species; (4) the few possible examples of the SB process do not increase adaptation in the way envisioned by Wright; (5) there are almost no empirical observations that are better explained by Wright's mechanism than by mass selection; and (6) because of the complexity of the SBT, it is impossible to test Wright's claim that it is a common evolutionary process. In view of these problems, we think that it is unreasonable to consider the SBT an important explanation of adaptation in nature.[26]

Now of course, nothing thus far proves the influence of Herbert Spencer. But even if we assume that only some of this is true, we are forced toward an asymmetrical position. If there were good reason to think that Wright's theory was conceptually clear and essentially true, then one might simply argue that – no matter how similar it seems to anything written by Spencer or anyone else – the reason for its taking the form that it does is that it corresponds to the way the world is. I do not need Spencer or anyone else to tell me that I am writing in English and that this sentence has a main verb. Why then pin an influence on Spencer? At least, one cannot say that there must be one. But since Wright's theory so clearly does not correspond to physical reality, one is driven to a search for sources. A photograph of Iowa cornfields is one thing. When Van Gogh gets out his pallette and brush, it is quite another. There had to be something that made Sewall Wright come up with what he did, and the way

---

[25] Jerry A. Coyne, Nicholas H. Barton, and Michael Turelli, "Perspective: A Critique of Sewall Wright's Shifting Balance Theory of Evolution," *Evolution* 51 (1997): 643–71, at pp. 643–4.
[26] Coyne et al., "Perspective," pp. 644–5.

the world turns is not it. Let us therefore turn our gaze backward, and do the most obvious thing. What kind of theory does Wright's look like? If you were looking for influences, what does it remind you of? The U.S. Supreme Court building looks like a Greek building rather than a Mexican building, so let us start with that. And similarly for the Shifting Balance Theory.

NATURAL SELECTION OR DYNAMIC EQUILIBRIUM?

Since the consensus is that it was Darwinism redux that happened at the beginning of the 1930s, let us first ask if Wright's theory looks like the theory of the *Origin of Species*. To this question, only one reply is possible: you must be kidding! Landscapes, peaks, drifting down, regrouping in new formations away from selection – the point at which the really important innovations are occurring – and then and only then selection in a backup, clean-up role, with groups within the whole population fighting it out: this may be many things, but it is not Charles Darwin. Fisher, a fanatical Darwinian, saw that right off. No one could make genetic drift as significant as did Wright and still be a Darwinian – especially since drift was not given a minor role but played *the* crucial part in evolutionary advance. More than this. You may object – what is certainly true – that although selection had a minor role in the early 1930s, at the end of the decade the empiricists (Dobzhansky particularly) found strong evidence of selection where once drift had been supposed. Hence, even though the early version of the theory was not very Darwinian, it was still potentially Darwinian, for it could be modified and selection could be given a bigger role. But in a way, this backfires, even though – precisely though – Wright did bring in more selection to his theory. He himself was not that bothered about the change – more accurately, he himself was supremely indifferent to the change – because for him the details of the mechanism were simply not that important. What counted was the overall picture. Think for a moment about the name of Wright's theory, something that often puzzles people: the Shifting Balance Theory of evolution. What is balanced? What is shifting? Wright is explicit on this. We have a balance – might one say an equilibrium? – and then for various imposed reasons, this gets destabilized. Then we get a move – a shift – to a new position. What is balanced? They are the forces tending to similarity and those tending to difference – those forces making for genetic homogeneity and those making for genetic heterogeneity. This is Wright's language, not mine.

Evolution as a process of cumulative change depends on a proper balance of the conditions, which, at each level of organization – gene, chromosome, cell individual, local race – make for genetic homogeneity or genetic heterogeneity of the species. . . . The type and rate of evolution in such a system depend on the balance among the evolutionary pressures considered here.[27]

By this time – and if nothing else is twigging you, the hierarchical language should – the case is almost overwhelming that with the Shifting Balance Theory we are looking, not at something Darwinian, but at something very Spencerian. We have stability; we have disruption; we have a vital non-Darwinian shift that creates new innovations; we have a return to stability. We have, to use Wright's own language, a tension between "homogeneity" and "heterogeneity." We have, to use a phrase, dynamic equilibrium. Of course, the two positions – Wright's and Spencer's – are not identical. Wright has genes; Spencer does not. Spencer is a Lamarckian; Wright is not. But without denying these differences, in a sense they are trivial, because for both men what really counts is the big picture. The details of the mechanisms can be filled in around this picture. Both are prepared to use selection, but for neither is it the be-all and end-all – as it was for Darwin and Fisher.

But what about the all-important question of progress? Does not the case for similarity come tumbling right down here? Spencer was the ultimate progressionist. Wright has not a mention of it in his theory. This surely divides them. It is at this point that we have to go back to the matter of professionalism. Genetics was an insecure subject at the beginning of the last century. For ten years, Wright worked in the USDA – and (speaking now with the authority of thirty-five years of teaching at an agricultural college) everyone knows that, in the academic pecking order, agriculture rates just above education, and below even sociology. Then, when he got a faculty position, Wright was low on the status totem pole. Amazingly to us today, back in those days genetics came below embryology. And evolution was even more insecure. It simply had to be presented without a whiff of philosophy or religion or whatever. Wright knew that and admitted that, even in his papers. He had to stay away from "speculation." And right at the top of the maxi-to-be-avoideds would be speculation about progress and the triumph of humans, the Anglo-Saxon humans in particular. But this does not mean that progress was not there in Wright's work. It does not mean that progress was not there, in an absolutely fundamental Spencerian fashion (rather than in a Darwinian add-on fashion). For a progressionism booster, who was also desperate to be seen as a professionalism booster, the best kind of theory would be one that had no necessary progress built in, but that would lend itself very readily to a progress-impregnated interpretation – indeed, that would beg for a

---

[27] Wright, "Mendelian," p. 158

progress-impregnated interpretation. And this, of course, was precisely the Shifting Balance Theory! The landscape could be like a waterbed. As one peak goes up another goes down, and ultimately you are right back where you started. Or it could be like the Himalayas, where things are pretty much fixed in rock, and Mount Everest is not about to sink below its neighbors. So no one could accuse Wright of being a progressionist with respect to the theory.

But progress was there for the taking if you wanted it. And Wright wanted it. Progress is right there in the first little pre-paper sketch that Wright sent to Fisher. And it is there from then on. Wright had some very strange metaphysical beliefs about everything – everything! – having consciousness, from molecules to men; he believed that perhaps we are on the way up to a kind of super organism, with super consciousness. This "panpsychic monism" is deeply progressionist, hierarchically and temporally.

The greatest difficulty is in appreciating the possibility of the integration of many largely isolated minds into a higher unitary field of consciousness such as must necessarily occur under this viewpoint in the organism in relation to its cells; in these in relation to their molecules and in these in relation to their molecules and these in relation to more ultimate entities. The observable hierarchy of physical organization must be the external aspect of a hierarchy of mind.[28]

So what I am concluding is that with respect to progress, as with respect to much else, Wright's is just the kind of theory that a 1930s Spencerian evolutionist, with aspirations to professionalism, would be expected to produce.

Let me add a couple of historical footnotes by way of backing. First, everyone after Wright – from Dobzhansky on – interpreted the adaptive landscape scenario in a progressivist fashion. Mayr, Stebbins, and G. G. Simpson (the author of *Tempo and Mode in Evolution* and the paleontologist of the group) were all ardent progressionists and took Wright's theory as their starting point. So the progressionism is a figment of more imaginations than mine. Second, in line with his indifference to the rising flood of selectionism, Wright basically was not that bothered about evolution-as-a-science after he had published his theory. It is true that, in the late 1930s and early 1940s, Wright wrote some fundamental papers with Dobzhansky, but the impetus came from Dobzhansky, and Wright got out of the collaboration as soon as he could. He never worked on evolution himself, devoting his energies to increasingly dated genetical studies with guinea pigs. He supervised just a couple of evolutionary theses out of over thirty in all, and he simply did not want to talk about

---

[28] Sewall Wright to J. T. McNeill, November 12, 1943, in the Sewall Wright Papers at the American Philosophical Society, Philadelphia.

evolution with his students. But he loved to talk about it with philosophers and theologians, especially his pals at the University of Chicago Faculty Club. This is really odd at the best of times, and it is truly odd without the missing factor of Spencer's influence.[29]

SOURCES

We come to the final part of the discussion on Wright. Is there any direct evidence of a Spencerian input? I am certainly not going to say that Spencer was the only input. Apart from all the genetics, Provine has shown (what Wright himself acknowledged) that the time at the USDA was crucially important. In particular, Wright did a massive analytic study of shorthorn cattle, and this convinced him that selection could not work in large groups – the secret is breaking the population into small isolated numbers, trying to effect change first in them, and only later returning to the large group. This was undoubtedly built into the Shifting Balance Theory. But this in itself did not make for a theory of evolution, and so we start to look farther afield. One influence that Wright acknowledged was the French philosopher Henri Bergson, the author of the vitalist classic *Creative Evolution*; and one can certainly see traces of this. Bergson was no great enthusiast for selection, thinking that it did not solve the problem of new and innovative features – the very things that Wright highlighted as needing more than mere selection. More than this. There is in Bergson a hint of the adaptive landscape metaphor. And Bergson was an ardent progressionist – although, showing that we should not take Bergson as squeezing out Spencer, we should note that Bergson himself always acknowledged the importance of Spencer's work in his own thinking, as something that directed his own thought on evolution.

Another apparent influence on Wright was a now-unknown chemist from Liverpool University in England, one Benjamin Moore. He wrote in terms that almost seem to have been cribbed from the Synthetic Philosophy.

It is only necessary for the atomic basis to our chemistry to realize that the atom, just like the chemical molecule at the different stage, or the fixed organic species of the biologist, is a point of the stable equilibrium in upward evolution. Between each two such points there lies a region of unstable equilibrium, and as matter becomes more

---

[29] I discuss this in *Monad to Man*. Two key pieces of information were letters to me – one from Janice Spofford, Wright's last doctoral student, of August 10, 1995; and the other from James Crow, the distinguished population geneticist and longtime friend and colleague of Wright, who retired from Chicago and went to Wisconsin for over thirty years (in the event, longer than his time as a full-time faculty member at Chicago!), of August 14, 1995.

charged with energy, surging and transformations occur, and in the greater number of cases when the cycle is complete, the matter drops back again to its stable point. But occasionally when a supply of energy at high-potential, or concentration, is available, there is a huge wave of uplifting which carries the matter involved over a hill crest into a higher hollow of stable equilibrium, and a new type of matter becomes evolved at the expense of kinetic energy passing over into latent energy or potentia.[30]

Then, finally, we start to corner in on Spencer himself. There is the home and early background. Sewall's father was an economist who taught his own son as an undergraduate, and who later was on the Harvard faculty when the son was a graduate student, and who apparently was much given to progressivist-type thinking in social and other spheres. Sewall's first teacher of biology, Wilhelmine Marie Entemann (Key), was apparently a Spencer enthusiast and had herself been educated by followers of Spencer – her doctoral work was supervised by Charles Otis Whitman, an explicit enthusiast for dynamic equilibrium thinking. Sewall and his brother Quincy – the latter a specialist in international law and also to become (like Sewall) a professor at Chicago – corresponded knowledgeably about Spencer, and Sewall (as a student) apparently had a picture of Spencer (Darwin also) on his wall.[31] And then, above all, there was the influence at Harvard of the chemist (and Sewall's teacher) L. J. Henderson, the author of the well-known work *The Fitness of the Environment* and a Spencer fanatic. Running through Henderson's writings are all sorts of organismic analogies, movements upward, changes from simplicity to complexity, and most prominently, that ever-changing flow to and from a state of balance. As Henderson said explicitly: "Spencer's belief in the tendency toward dynamic equilibrium in all things is of course fully justified."[32]

The student was brought under the spell – "I was always very much impressed with Henderson's ideas"[33] – and acknowledged explicitly the direct influence back to Spencer – "I found him a very stimulating lecturer and got lots of ideas from him, 'condition of dynamic equilibrium' etc."[34] And the young thinker worked things out, particularly in letters to Quincy. The organismic analogy:

Thus the body is not an absolute monarchy in which the bulk of the cells are mere mechanisms, directed in every action by a central unit. It is democracy or perhaps better is limited monarchy. In the main each part knows what to do and does it of its

[30] B. Moore, *The Origin and Nature of Life* (London: Williams and Norgate, 1913), p. 40.
[31] Sewall Wright to Quincy Wright, December 14, 1915, in Quincy Wright Papers, University of Chicago. The letters referred to in the following notes are in the same archive.
[32] L. J. Henderson, *The Order of Nature* (Cambridge, MA: Harvard University Press, 1917), p. 138.
[33] Interview with William Provine, June 4, 1976, Wright Papers.
[34] Letter to Quincy Wright, January 10, 1916.

own accord, as occasion arises. Regulation from outside comes rather from suggestions from numerous peers, not in a single command from above.[35]

The hierarchical thinking, linking evolution and equilibrium:

My original idea was to classify all sciences by the unit of organization – electron, atom, animal, etc. – with which they deal subdividing on a fourfold basis –

A. Condition in equilibrium
  1. Description of organization
  2. Mechanism of maintenance of equil.
B. Change of equilibrium (Evolution)
  3. Description of changes (history)
  4. Mechanism of change[36]

And another attempt, again linking evolution and equilibrium:

The difficulty of classification is well illustrated by my own science, genetics – from one point of view it deals with the organization of the cell and has very close relations with cytology, then it deals with the mechanism of individual development – the mode in which developmental factors are represented in the one cell stage, – and finally it deals with both the maintenance of equilibrium in the species (heredity) but also the mechanisms of change in this equilibrium (variation by recombination of factors and otherwise).[37]

And wrapping everything up in terms of progress:

Darwinists would hold that the most rapid evolution would follow from a happy mean between conditions which permit the existence of a wide range of variations, – many of them more or less injurious – which can recombine in all possible ways – and conditions which tend to eliminate the more unfit. To use a human analogy, we do not expect civilization to advance most rapidly either in the arctic zone where existence depends on following one very definite mode of life or in the tropics where conditions of life are too easy. . . . The greatest progress should result in a society which is neither crystallized into a caste system nor so fluid that individuals of a family, which has produced favorable variations and done much for progress in the past, receive no advantage over inferior families. The problem of statesmanship is to adjust laws so that there is just the degree of viscosity in all respects which gives the maximum progress. It is a problem of maxima and minima and therefore much more difficult than progress toward an absolute democratic or absolute aristocratic ideal.[38]

If all this does not add up to a smoking gun, I don't know what does. The Shifting Balance Theory – the genesis of which, incidentally, apparently goes back before 1930 (apparently it was first written up around the time Wright went to Chicago, that is, around 1925) – was Herbert Spencer updated. There

---

[35] Letter to Quincy Wright, December 14, 1915.
[36] Letter to Quincy Wright, February 27, 1916.
[37] Ibid.   [38] Letter to Quincy Wright, October 17, 1915.

are links back through Henderson and his influence to the Synthetic Theorist himself. R. A. Fisher was a Darwinian. Sewall Wright was not.[39]

CONCLUSION

True to my promise, I will leave things here. I will forbear mentioning that the Stephen Jay Gould/Niles Eldredge non-Darwinian, paleontological theory of punctuated equilibrium sounds very much like a Spencerian offshoot to me (especially in Gould's final massive testament, *The Structure of Evolutionary Theory*).[40] I will leave unsaid the fact that the arch-progressionist of American evolutionary biology today, Edward O. Wilson, has on his wall a picture of Herbert Spencer, whom he much admires. ("Great man, Mike! Great man!") Debunking one myth per paper is enough heresy even for me.

[39] Timing, of course, is everything. I have stressed toward the end of this discussion that I do not see Spencer as the only influence on Wright, but I do see him as a major one – indeed, if one allows the transmission of ideas through others (as through Henderson), then my claim is that Spencer is far and away the major influence. What about the most obvious of all philosophical influences, namely the American pragmatists? Wright was a student at Harvard and a faculty member at Chicago, so a link seems at least plausible, and I am certainly not going to deny that there was such a link. Indeed, qua mathematician, as opposed to Fisher, Wright seems very much to be a pragmatist – concerned with results rather than elegance – if not a pragmatist in the philosophical sense. It is just that I have not seen any connections – Sewall and Quincy talk about Spencer and not about pragmatism – and certainly if Provine's claim that the SBT was essentially complete by the time Wright got to Chicago, we can discount one obvious source.

[40] Stephen J. Gould, *The Structure of Evolutionary Theory* (Cambridge, MA: Harvard University Press, 2002).

CHAPTER EIGHT

"The Ninth Mortal Sin"
The Lamarckism of W. M. Wheeler

Charlotte Sleigh

INTRODUCTION

In a lecture to the philosophical Royce Club of Harvard in 1917, William Morton Wheeler jovially referred to Lamarckism as "the ninth mortal sin."[1] Wheeler (1865–1937) was by then the world's leading figure in myrmecology (the study of ants), and there was a serious comment underlying his remark; something about his practice of myrmecology had placed him outside the orthodoxy of biology. A sketch of his career hints at why this might be so. Wheeler had formalized an early interest in natural history when he went to Chicago to study embryology under C. O. Whitman. Moving to Texas some time later, he became intrigued by the local ant fauna and returned to his earlier natural-historical passions. The fascination with myrmecology lasted for the remaining forty or so years of his life, to the virtual exclusion of all other types of animal – humans excepted. Most of this time (1908–37) was spent at Harvard. A romantic natural historian at heart, Wheeler was uncomfortably lodged with the applied biologists in the graduate school; by preference he mixed with Cambridge's philosophers, psychologists, and sociologists, along with the zoologists in the Museum of Comparative Zoology.[2]

---

[1] William M. Wheeler, "On Instincts," *Journal of Abnormal Psychology* 15 (1921): 295–318, at p. 303.
[2] Wheeler defines himself as a romantic natural historian in William M. Wheeler, "A Notable Contribution to Entomology," *Quarterly Review of Biology* 11 (1936): 337–41, at pp. 340–1.

151

Ants had long been a riddle to evolutionists like Wheeler; though individually simple, their group behavior was extremely complex, embracing activities
that were variously interpreted as fungus farming, aphid farming, slave making, mutual feeding, brood nursing, nest building, caste creation, and warfare,
not to mention their manifold ecological relationships with other species, both
within and without the nest. Their unorthodox family groupings, and consequently their unusual patterns of inheritance, served only to deepen the
riddle. These issues are discussed in the first part of this chapter.

Yet Wheeler paid scarcely any attention to the seemingly intractable issues
of acquired inheritance among sterile castes; the remainder of the chapter
engages instead with his distinctive approach to ants, asking in what sense
it *was*, then, that Wheeler considered himself a Lamarckian. The answer to
this question falls into three parts. First, Wheeler had a profound conviction
of the priority of function over structure, expressing from very early on in
his career a metaphysical skepticism about the integrity even of the "individual" organism. Second, his commitment to natural history led him to seek a
new intellectual forebear in place of the laboratory workers' Darwin. George
Orwell once described Jonathan Swift as a "perverse Tory," meaning that the
Dean aligned himself with the party only because of his frustration with the
alternative Whig position and despite the fact that he did not subscribe to
many of the Tories' traditional views. I believe that the description "perverse
Lamarckian" would suit Wheeler rather well, for he was searching for means
by which to distinguish his historical evolutionary work from the atemporal
laboratory efforts of his peers, which he rejected. Despite embracing the title
"Lamarckian," Wheeler did not blindly seek to demonstrate a crude version
of acquired inheritance; rather, it was the best label to suit his conviction
that evolution was a more holistic and reflexive process than one involving
simple individuals. Third, Lamarck provided Wheeler with a fully naturalized
eugenic imperative. Effacing the metaphysical rupture between natural and
artificial selection, eugenics was for him simply the newest expression of the
cyclic psychophysical development of life. Though these perspectives were
not unique in themselves, together they explain what Wheeler meant when
he chose to align himself with Lamarck.

I. DARWIN AND THE LAMARCKIAN "PROBLEM" OF ANTS

For Darwin, the behavior of insects was so striking – "ranked by naturalists as
the most wonderful of all known instincts" – that it was one of the principal
things for which any theory of evolution had to account. In his chapter on
the subject in the *Origin of Species*, he aimed to show that even the apparently

complex instincts of slave making among ants, and of hive making among bees, might be accounted for by the gradual variation and selection of simple instincts, in just the same way as physical traits evolved.[3]

However, in addition to the mechanism of selection, a tempting analogy between instinct and habit presented itself to Darwin. It seemed to Darwin that habits *might* be inherited and thus to all intents and purposes become indistinguishable from instinct. In one passage, Darwin appears to reject this possibility out of hand:

> If we suppose any habitual action to become inherited – and I think it can be shown that this does sometimes happen – then the resemblance between what originally was a habit and an instinct becomes so close as not to be distinguished. . . . But it would be the most serious error to suppose that the greater number of instincts have been acquired by habit in one generation, and then transmitted by inheritance to succeeding generations. It can be clearly shown that the most wonderful instincts with which we are acquainted, namely, those of the hive-bee and of many ants, could not possibly have been thus acquired.[4]

Yet a certain hesitation over the relation between these two models of instinct formation is nevertheless detectable even in the first edition of the *Origin*. Despite his rejection of habit-based instinct acquisition, Darwin considered that the "domestic" instincts of dogs (such as the instinct of a shepherd dog to run around a flock) were less fixed. This was because these instincts had been acted on by weaker (i.e., artificial) selective forces and because they had been transmitted through relatively few generations. In other words, training had become fixed in the same kind of way as true instinct, though less strongly. Ostensibly, Darwin carefully steered away from discussing "the origin of the primary mental powers" in 1859, just as he had from the origin "of life itself." It can clearly be seen that he did not entirely succeed; indeed, in later life Darwin came to embrace acquired characteristics more whole-heartedly.[5] It is important to bear in mind that the modern conception of Darwinism – that is, strict selectionism – was not a consensus model until

---

[3] Charles Darwin, *On the Origin of Species* (Oxford: Oxford University Press, 1996 [1859]), pp. 169–98. For a summary, see Roger Smith, *The Fontana History of the Human Sciences* (London: HarperCollins, 1997), pp. 462–3, 472–3.

[4] Darwin, *Origin of Species*, p. 170.

[5] See Robert J. Richards, *Darwin and the Emergence of Evolutionary Theories of Mind and Behavior* (London: University of Chicago Press, 1987), pp. 90–8, for Darwin's early theory of habit-instinct adaptation. Although it is interesting to see how Darwin's theory is similar to those of Forel and others (discussed later), Forel and other entomologists made it their own, working on it at length and establishing it as a guiding scientific norm. See Charlotte Sleigh, *Six Legs Better* (Johns Hopkins University Press, forthcoming).

the late 1930s.[6] To most people, it seemed that some combination of selection and character acquisition was responsible for evolutionary change.

Curiously, although Darwin used the example of infertile worker ants to "disprove" Lamarckism in the *Origin* (because sterile, they could not pass on any characteristics which they had acquired), virtually all early twentieth-century entomologists were convinced that some kind of character acquisition was the best way, or indeed the only way, to explain the complex behavior of ants.[7] How, then, did myrmecologists maintain their Lamarckian stance? By way of solution, some commentators suggested that sterility had evolved only *after* the phyletic development of caste differentiation. Wheeler, however, dismissed this claim, as it would entail a history in which formicaries had been formed by voluntary associations of adult insects – a claim that he and most other myrmecologists could not accept.[8] Wheeler's own response to the anti-Lamarckian problem was to point out that there were significantly more fertile worker ants than were generally supposed. When the queen was removed from the nest, fertile workers were especially likely to appear in significant number. Thus there was, for the purposes of neo-Lamarckian evolution, enough potential for the characters of the workers to be fed back into a new generation.

If we grant the possibility of a periodical influx of worker germ-plasm into that of the species, the transmission of characters acquired by this caste is no more impossible than it is in other animals, and the social insects should no longer be considered as furnishing conclusive proof of Weismannism.[9]

---

[6] Peter Bowler, *The Eclipse of Darwinism: Anti-Darwinian Evolutionary Theories in the Decades around 1900* (Baltimore: Johns Hopkins University Press, 1983).

[7] On Darwinism and Lamarckism, see Bowler, *Eclipse of Darwinism*; on Lamarck's mental theories, see Richards, *Darwin and the Emergence of Evolutionary Theories*, pp. 47–57. The confusion between the various theories of evolution is illustrated by E. C. Wilm, *The Theories of Instinct: A Study in the History of Psychology* (New Haven, CT: Yale University Press / London: Oxford University Press, 1925), which bizarrely describes Forel as a neo-Weismannian – and this in a book that had been read over by Wheeler. (Presumably, Wheeler was having cold feet about strict Lamarckism at the time, or else wished to distinguish himself from his myrmecological senior. Alternatively, he may have been referring to Forel's position on the Weismann/Boveri debate.)

[8] William M. Wheeler, *Ants: Their Structure, Development and Behavior* (New York: Columbia University Press, 1910), pp. 116–17. The claim that Wheeler dismissed was very similar to Herbert Spencer's theory on the origin of ant consociations.

[9] Wheeler, *Ants*, pp. 115–16. Erich Wasmann and Adele M. Fielde had also found numerous fertile workers in the formicary. Paul Marchal, "La Reproduction et l'Evolution des Guêpes Sociales," *Archives de Zoologie Expérimentale et Générale*, third series, 3 (1896): 1–100, showed that young sterile workers could be induced to become fertile through overnourishment. E. L. Bouvier, in *Le Communisme chez les Insectes* (Paris: Flammarion, 1926), judged their appearance to be normal under advantageous circumstances for the nest as a whole.

Wheeler considered that Weismannian theory did not go any distance toward answering questions about the origins of ant polymorphism (i.e., their different caste forms). In his 1910 masterwork *Ants*, Wheeler dismissed Weismann as completely discredited. It is hard to tell whether this was bravado or a genuinely held conviction on Wheeler's part, but what is undeniable is that the exact mechanism of evolutionary feedback was not of particular interest to him; the discussion of the Weismann question takes less than two pages in a book of nearly seven hundred. As a naturalist, Wheeler was unimpressed by Weismann's theoretical approach, dismissing it as only a restatement of the question: a mere "photograph" of the problem of polymorphism.[10] The development of worker ants might indeed be due to Weismannian germinal predetermination of their eggs, Wheeler admitted, but the key question was rather *how did that predetermination arise?* The conclusion he reached, along with the Italian zoologist and myrmecologist Carlo Emery, was that a subtle deflection of the germ-plasm along caste-specific developmental pathways was achieved by differential feeding of the larvae, a "decision" made unconsciously by the nest as a whole. In his 1917 paper "On Instincts," Wheeler repeats the same point:

[O]ur inability to detect the inheritance of an acquired character is probably due to the fact that its visible appearance is preceded in phylogeny by a period of many generations during which it is inherited only as a function associated with alterations of structure too subtle to be revealed by our present very crude methods of observation and experiment. Mendelian ratios would therefore be merely the method of inheritance of the stereotyped end-products of a long evolution and would not represent the actual phylogenetic method of the development of such characters.[11]

Wheeler was unfazed by the lack of experimental evidence for the acquisition of acquired characteristics. It was obvious to him that what was needed was an imaginative reconstruction of phylogenetic time, a reconstruction that could be achieved only through sophistication in natural history.

Thus Wheeler did not spend very much time addressing the so-called Lamarckian problem. It is now common knowledge that there were a variety of disputed "Darwinisms" in the late nineteenth century; here we have a concrete example of an entire discipline founded, almost unproblematically, upon non-Darwinian theory. Although Kuhn's paradigm theory is invoked too often and too loosely, this may actually be a rather good example of paradigmatic incommensurability. Wheeler simply did not think that the questions and model solutions of genetic biology or Weismannism were relevant to his work. I have done my best here to draw out myrmecologists' explicit responses

[10] Wheeler, *Ants*, p. 100.   [11] Wheeler, "On Instincts," p. 303.

to the Weismannian debate, but these points were not a priority for them. Having dismissed the negative question about Lamarckism's opposition, I shall now turn to the positive question: the tradition within which we may locate Wheeler.

II. STRUCTURE AND FUNCTION

In 1874, the Swiss psychiatrist and neurologist Auguste Forel claimed to have united, for the first time, the behavioral and taxonomic study of ants.[12] Forel's studies set the pace for myrmecology in the late nineteenth century and established as standard a now strange-sounding subdiscipline: insect psychology. Forel was a committed monist, and this affected his work with both ants and humans. In the case of humans, Forel firmly believed that mind and brain were one and the same, a conviction validating his reeducation of hysterics, alcoholics, and other patients: he was literally reconstructing their brains. Ants were a simpler model of the same processes. Their phylogenetic history was the grand theme echoed briefly in the ontogenetic history of the individual human; the behavioral adjustments of ants to changing conditions had become engraved in the heritable structures of their nervous systems.

Forel's theories about ant psychology drew on a recent efflorescence of work on "organic memory." A pivotal piece regarding the inheritance of the psyche was Ewald Hering's 1870 lecture to the Viennese Imperial Academy of Science, "Über das Gedächtnis als eine allgemeine Funktion von Organisierten Materie" (On memory as a general function of organized matter). Hering's theory that all living matter had the capacity to reproduce memory was respectfully cited into the twentieth century – by Freud, among others.[13] In Britain, Hering's ideas were enthusiastically taken up by Samuel Butler, who had independently worked out something very similar.[14] In France, meanwhile, the philosophical psychologist Théodule Ribot strove to incorporate memory into the fields of biology and physiological psychology and to turn it into a hereditary phenomenon.[15] The nationalist German biologist Ricard

[12] Auguste Forel, *Les Fourmis de la Suisse* (Basle, Geneva, Lyon: H. Georg, 1874), p. i.
[13] Laura Otis, *Organic Memory: History and the Body in the Late Nineteenth and Early Twentieth Centuries* (Lincoln, and London: University of Nebraska Press, 1994), pp. 10–14.
[14] See Peter Raby, *Samuel Butler* (London: Hogarth, 1991); Samuel Butler, *Unconscious Memory* (London: Jonathan Cape, 1920 [1880]); Butler, *Evolution Old and New: Or, the Theories of Buffon, Dr. Erasmus Darwin, and Lamarck, as Compared with that of Mr. Charles Darwin* (London: Hardwicke and Bogue, 1879); Butler, *Life and Habit* (London: Trubner, 1878).
[15] Otis, *Organic Memory*, pp. 14–17. Ribot collaborated with Alfred Espinas in his translation of Herbert Spencer's work. Espinas was influential among myrmecologists because of his book *Des Sociétés Animales* (Paris: Germer Ballière, 1878 [1877]).

Semon was responsible for the theory of the "mneme," which explained how mental phenomena became associated with one another as memories over the course of a lifetime and how they were coordinated over the course of phyletic time as heredity.[16] Organic memory was an important basis for science across Europe.

Forel's particular organic memory theory used a mechanism developed in the context of his neurological experience.[17] In a novel situation, the senses of an organism would receive certain stimuli. After this had ceased, a permanent change resulted in the nervous system of the organism (usually the brain), called the "engram." With repeated presentation of the stimulus, the engramic response could eventually be elicited even by a weakened form of that stimulus. This state corresponded to the psychological condition of association. Sometimes there would be a conflict between internal prompt and external stimulus; this was resolved within the life of the organism by neuronal "regeneration." Forel's business as a psychiatrist was to help his patients to relearn such habitual behavior.

In the long term, these associative behaviors became fixed as innate instincts. Moreover, the race would acquire hereditary physiological conditions enabling it to react to its environment in a manner complementary with the instinct. Termites provided Forel with a particularly clear example of this psychophysical complementarity; soldiers with immense armoured heads blocked the entrance to the nest, repelling would-be intruders. These species had gradually ceased the practice of blocking the entrance with gathered or secreted materials and now accomplished the task with a specialized caste.[18] It made no sense to ask which came first, the behavior (useless without the large head) or the physiology (mere monstrosity without the behavior). The two were acquired in tandem, anatomy following behavioral changes. Hence a functional discussion of anatomy (especially in its connection to

---

[16] Ricard Semon, *The Mneme* (London: Ruskin New York: Macmillan, 1921 [1904]). Wheeler had taken notes on E. Rignano, *Upon the Inheritance of Acquired Characteristics*, trans. B. C. M. Harvey (Chicago: University of Chicago Press, 1911), in which he explained teleology as due to "mnemic peculiarities." Notes taken by Wheeler, Wheeler Papers, HUGFP 87.65, Box 1, Pusey Library, Harvard.

[17] Auguste Forel, *Hypnotism, or Suggestion and Psychotherapy: A Study of the Psychological, Psycho-Physiological and Therapeutic Aspects of Hypnotism*, trans. H. W. Armit (London: Rebman, 1906 [1889]), pp. 4–5; Auguste Forel, *The Social World of the Ants Compared with That of Man*, 2 vols., trans. C. K. Ogden (London: G. P. Putnam's Sons, 1928 [1921–22]), vol. 1, pp. 182–5. Forel acknowledged that he had borrowed his terminology from Ricard Semon. In Forel, *Hypnotism*, p. 5, he suggested that "engraphs [much the same as engrams] might serve as an explanation for De Vries' mutations."

[18] Edouard Bugnion, "The Origin of Instinct," in Forel, *Social World of the Ants*, vol. 2, pp. 384–5. See also William M. Wheeler, *Foibles of Insects and Men* (New York: Knopf, 1928), pp. 37–43.

sensation) was the key to understanding mind. The senses, the nerves, were a physical reflection of the evolved, adaptive aspects of the insect psyche.[19] This was clearly a quasi-Lamarckian theory.

Forel and his successors mixed physical and mental evolution in the organic memory tradition without explicitly dwelling on Lamarck. Forel described the infamous series of experiments on salamanders in which the neo-Lamarckian Paul Kammerer claimed to have provoked the heritable acquisition of novel skin coloration, but he did not seem particularly convinced of their significance.[20] Forel found himself unable to account satisfactorily for the phylogenetic (heritable) incorporation of engrams. Yet time and again, he insisted that however perplexing the problem of inherited instinct might appear, he could do no better than to quote Hering's aphorism, "Instinct is the memory of the species."[21]

A large number of entomologists based their studies in insect psychology along similar quasi-Lamarckian lines, sharing the conviction that function took priority over structure. Their vision of evolution was one of a psychophysical economy as the motor of change.[22] Individuals changed their behavior in order to act most efficiently under any reasonably permanent new circumstances, and these changes became fixed in future generations. The French entomologists Eugène Bouvier (whom Wheeler described somewhat mischievously as "a sane and catholic Neolamarckian")[23] and Emile Roubaud used this as a basis for their work. Alfred Giard, a French contemporary of Forel's, was very influential in his Lamarckism and sponsored a younger generation of students, including Georges Bohn and Henri Piéron. Quasi-Lamarckian work on insect psychology was also supported by French institutions, notably the Institut Générale Psychologique. In Switzerland, Forel's brother-in-law Edouard Bugnion helped to establish his quasi-Lamarckian views, and the Swiss-born psychiatrist Rudolph Brun gave an authoritative reformulation of

---

[19] Forel, *Social World of the Ants*, vol. 1, p. 233. Forel focused on this point in his book *The Senses of Insects*, trans. Macleod Yearsley (London: Methuen, 1908 [1878–1906]). See also H. Eltringham, *The Senses of Insects* (London: Methuen, 1933).

[20] Forel, *Social World of the Ants*, vol. 1, pp. 16–19.

[21] Forel, *Social World of the Ants*, vol. 1, p. xliii.

[22] Charlotte Sleigh, "Brave New Worlds: Trophallaxis and the Origin of Society in the Early Twentieth Century," *Journal of the History of the Behavioral Sciences* 38 (2002): 133–56; Sleigh, *Six Legs Better*.

[23] William M. Wheeler, review of Bouvier, *La Vie Psychique des Insectes, Science* 52 (1920): 443–6, at p. 444. On French animal psychology around the turn of the twentieth century, see Marion Thomas, *Rethinking the History of Ethology: French Animal Behaviour Studies in the Third Republic (1870–1940)*, unpublished Ph.D. thesis, University of Manchester, 2003.

Forel's theories as applied to insect orientation. American quasi-Lamarckians included Philip Rau and Charles H. Turner; the U.S. entomologist Alphaeus S. Packard devoted his final major work to Lamarck.[24] Meanwhile, in Britain, the ethologist William Thorpe began his career as a quasi-Lamarckian entomologist. Forel's work, the inspiration for a generation of insect psychologists, thus forms an obvious route by which Wheeler had come to his position vis-à-vis Lamarck. Although Forel himself did not actively champion Lamarck, his focus on function and acquired characteristics were certainly complementary with Lamarckism.

There was, however, more to Wheeler's philosophy of biology than diluted Lamarckism. Wheeler, an exceptionally well-read and philosophical biologist, consciously considered a wide range of possible models as the basis for his myrmecology. He interpreted all of them in a manner compatible with his construction of Lamarck: the priority of function over structure.

In the early stages of his career, Wheeler was greatly influenced by Henri Bergson. As early as 1904, Wheeler was creating a holist definition that destabilized the integrity of the organism:

We may, in fact . . . take the point of view of the psychologist and the metaphysician rather than that of the morphologist. In other words, we may start with the behavior or the dynamic, i.e., physiological and psychological processes of the organism, and regard the structure as their result or objectivation [sic]. . . . In this sense the honeycomb is as much a part of the bee as her chitinous investment, and the nest is as much a part of the bird as her feathers, and every organism, as a living and acting being, fills a much greater sphere than that which is bounded by its integument.[25]

What biologists were pleased to call an "organism" was actually an unpredictable emergence from its "ethos," or, as Wheeler wrote in 1911, "neither a thing nor a concept, but a continual flux or process."[26] This focus on the

---

[24] Alphaeus S. Packard, *Lamarck: The Founder of Evolution; His Life and Work, with Translations of His Writings on Organic Evolution* (New York: Longmans, Green, 1901). See W. Conner Sorensen, *Brethren of the Net: American Entomology, 1840–1880* (Tuscaloosa and London: University of Alabama Press, 1995), pp. 197–213, on the evolutionary theory of American entomologists.

[25] William M. Wheeler, "Ethology and the Mutation Theory," *Science* 21 (1905): 535–40, at p. 535. This lecture was given in 1904 to the American Society of Naturalists. The definition of an organism that included its environment is closely related to the work of Jacob von Uexküll, on whom see Anne Harrington, *Reenchanted Science: Holism in German Culture from Wilhelm II to Hitler* (Princeton, NJ: Princeton University Press, 1996), pp. 34–71.

[26] The identity between the individual and the group organism in Wheeler's work cannot be overemphasized. Wheeler wrote that "the animal colony is a true organism and not merely the analogue of the person." (William M. Wheeler, "The Ant-Colony as an Organism," *Journal of Morphology* 22 (1911): 301–25, at p. 308.) For experimental purposes, the superorganism was

dynamic processes of biology, and the refusal of categorical stultification, constitutes an unmistakably Bergsonian outlook.

At the height of his infatuation with Bergson, Wheeler considered that the debate about the mechanism of evolution was an unnecessary semantic tangle. He refused to endorse Forel's theory, holding instead to a more holistic view of "becoming" in ants and humans:

The views on the origin of automatic behavior . . . are so diverse and conflicting that they cannot be satisfactorily considered without entering into a discussion of the doctrines of the Neodarwinians, Neolamarckians and those who believe in coincident, or organic selection. In my opinion we have little to gain at the present time from such a discussion. . . . It is, in fact, quite futile to attempt a phylogenetic derivation of the automatic from the plastic activities or vice versa, for both represent primitive and fundamental tendencies of living protoplasm, and hence of all organisms. As instinct, one of these tendencies reaches its most complex manifestation in the Formicidae, while the other blossoms in the intelligent activities of men.[27]

Wheeler outgrew Bergson by 1917 or so, but he retained an affinity for holistic and fluid thinking and throughout his life refused to be pinned down to reductionist definitions of the "organism."

A second strand of Wheeler's philosophy was his obsession with hierarchies and the related phenomenon of emergence. In his superorganism paper of 1910 (published in 1911), Wheeler explained that there was a whole hierarchy of "organisms" in nature. He began his scale with the conceptual biological unit of the "biophore," by which he meant the simplest possible organism. Biophores combined to form cells, which combined to form organisms, which themselves combined to form societies and thence "cœnobioses." Cœnobioses were "more or less definite consociations of animals and plants of different species" – something very like Alfred Tansley's concept of the ecosystem.[28] At

---

the ideal form of organism, since one could experiment with its component parts more easily than one could with an "ordinary" organism, without causing damage too great to obtain results.

[27] Wheeler, *Ants*, pp. 543–4.

[28] The "biophor" was posited by Ernst Brucker. The "Biocœnosis" was an important component of German *Heimat* philosophy, which incorporated nature sui generis and nature as human habitat. The *Heimat* concept inspired museum display techniques during the 1860s; insects were probably the first organisms to be displayed in cases at all of their life stages, among their natural vegetaion, and accompanied by their parasites. Susanne Koestering, "Visualising Biology: Ecology and *Heimat* in Natural History Displays in Germany, 1871–1914," paper given at conference of the International Society for the History, Philosophy and Social Studies of Biology, Oaxaca, Mexico, July 1999. The term "ecosystem" was coined by Tansley in 1935. See also Frank Golley, *A History of the Ecosystem Concept in Ecology* (New Haven, CT: Yale University Press, 1993).

each of these levels, there was a phenomenological emergence that was studied by a particular kind of scientist, from the geneticist to the ecologist.

The emergent nature of each successive level meant that no scientist could comment on a phenomenological level above his own, since its properties could not be predicted from his own level of study. Weismann, who had adopted the term "biophor" to denote the smallest autonomous component of the chromosome, was therefore stuck at the bottom of Wheeler's scientific heap. Wheeler, a student of the interspecific relationships of the ants' nest, was at the top of the scientific hierarchy and surveyed the whole of nature, right down to Weismann's level.

Wheeler's expertise in social evolution and emergence also gave him, he implicitly claimed, a unique familiarity with each level in its own right, for he construed each successive emergence as intrinsically social. First, each member of the hierarchy above biophores was an aggregation of the level below. Second, each animal was inherently social inasmuch as it existed as a part of the cœnobiosis. A third and final reason to regard all animals as social was that every animal had to seek out at least one moment of intimate social union in reproduction. Wheeler thus saw social phenomena as the primary units of nature; the reductive Weismannian did not study nature in its essentially social state.

Besides validating social insects as objects of study metaphysically more worthwhile than Weismann's, Wheeler's social hierarchies also challenged the selective basis of neo-Darwinism. Because all life was interrelated, competition between individuals in any meaningful sense was essentially absent. This was a decidedly non-Darwinian perspective. To be sure, Darwin had blurred the categories of variety and species in his own work. But Wheeler went further, eschewing the individual conflict central to a selectionist account of evolution. Wheeler was very explicit about his non-Darwinian stance, putting it in terms of the types of behavior – cooperative and individualistic – central to the two respective explanations of evolution. He designated individualism as "natural" within a Darwinian scheme of explanation, rendering a special explanation necessary for cooperation whenever it was observed. However, the study of complex interrelationships within nature persuaded Wheeler that the ordering of normal and out-of-the-ordinary ought to be reversed: that cooperation was the norm and selfish individualism – such as was exhibited by humans – a noteworthy phenomenon that needed to be accounted for when it was observed.[29] As late as 1923, five years after he professed to

---

[29] William M. Wheeler, *The Social Insects: Their Origin and Evolution* (London: Kegan Paul, Trench, Trubner, 1928), p. 5. In 1911, Wheeler claimed: "One of the fundamental tendencies

have left Bergson behind, Wheeler was still claiming that the whole of the organic realm constituted "one vast, living symplasm," whose fragmented parts formed one metaphysical whole of "Common Life." Because there was a blurring of these entities, competition among them was meaningless. Victorian notions of struggle were at most half the story of evolution, according to Wheeler, and cooperation the more significant factor.[30]

Toward the end of his life, Wheeler coedited a volume containing some of Lamarck's lesser-known manuscripts, most of which had been in Alexander Agassiz's collection and had remained at Harvard. Wheeler commented in his Preface that "after a long and unmerited neglect, at least in America and England, Lamarck is now held in . . . high esteem by the majority of biologists."[31] Though perhaps a little too sanguine, Wheeler's remark nevertheless indicates Lamarck's standing among his fellow entomologists and, not least, the strength of Wheeler's own feelings. Wheeler's Lamarckian heresy was no heresy at all among his psycho-entomological colleagues. Furthermore, it was strongly connected, via Forel, to a tradition of human psychology and psychoanalysis that interiorized Haeckel's principle of recapitulation, governed by a teleological drive of energy conservation. In Freud's consulting room, dangerously pent-up charges from childhood and primitivism were released, while in Wheeler's hunting grounds, such forces were constructed as the non-Darwinian force of evolution, molding first the behavioral and then the neural. Wheeler's stance *was* heretical inasmuch as it dispensed with competition altogether, making communal life the norm of nature.

III. BIOLOGY VERSUS HISTORY

Wheeler was repeatedly provoked to defend the value of natural history by feelings of professional threat from successful, academic, laboratory biologists.

In the scientific literature of the present time . . . natural history is so rarely mentioned that it seems to be the name of some extinct science, like alchemy or astrology. The term "naturalist" has also passed out of use. A few years ago, I was introduced to an audience by an eminent paleontologist as one of the last surviving naturalists, and, of course,

of life is sociogenic. Every organism manifests a strong predilection for seeking out other organisms and either assimilating them or co-operating with them to form a more comprehensive and efficient individual." Wheeler, "The Ant-Colony as an Organism," p. 324.
[30] William M. Wheeler, *Social Life among the Insects* (New York: Harcourt Brace, 1923), pp. 3–5.
[31] William M. Wheeler and Thomas Barbour, eds., *The Lamarck Manuscripts at Harvard* (Cambridge, MA: Harvard University Press, 1933), p. vii.

the audience eyed me as if it were catching its last glimpse of a living Brontosaurus. . . . I felt like the curator who overheard a little girl say, while she was being conducted through his zoological museum "Why, mother, this is a dead circus!"[32]

Whether or not it made him a fossil, Wheeler was unashamedly a naturalist. He was always a keen supporter of the Boston Society of Natural History and retained a special loyalty to the American Society of Naturalists throughout his career, using its meetings as a platform to present some of his most important papers.[33] He also defended natural history in general scientific circles, speaking in its favor on a number of occasions at the American Association for the Advancement of Science (AAAS). From very early in the twentieth century until his death, Wheeler consistently used his publications to reinforce the value of natural history.

In 1922, Wheeler exchanged letters with T. H. Morgan on the topic of saving naturalists, whom the bigger organizations – specifically, the zoological societies – were trying to drive to the wall. Morgan, the pioneer geneticist of the fruitfly, unsurprisingly suggested the incorporation of genetics into natural history, in order to make it more professionally palatable, and, facetiously, recommended holding the naturalists' meetings elsewhere, as far out of the way as possible, since such absence was bound to make the zoologists' hearts grow fonder.[34] In response to such attitudes, Wheeler hoped to defend "natural history" as a distinctive alternative program for biological research. In publishing his "Ant-Colony as an Organism" paper in the *Journal of Morphology*, he was arguably trying to place natural history and its objects of study on a par with laboratory science. And at the AAAS in 1928, Wheeler explicitly argued that the future of biological theory lay with "that union of historic and naturalistic interests which seems to inspire an ever-increasing number of our biologists and promises the fullest ultimate understanding of ultimate nature."[35]

[32] William M. Wheeler, "What is Natural History?," *Bulletin of the Boston Society of Natural History* 59 (1931): 3–12, at p. 4. See also Sally G. Kohlstedt, "From Learned Society to Public Museum: The Boston Society of Natural History," in Alexandra Oleson and John Voss, eds., *The Organization of Knowledge in Modern America, 1860–1920* (Baltimore and London: Johns Hopkins University Press, 1979), pp. 386–406. Kohlstedt describes the BSNH's efforts to redefine its identity and purposes as the institutions of science changed around it.

[33] For example, William M. Wheeler, "The Termitodoxa, or Biology and Society," *Scientific Monthly* 10 (1920): 113–24 (given as a talk in 1919); Wheeler, "Animal Societies: Biology and Society," *Scientific Monthly* 39 (1934): 289–301 (given as a talk in 1933).

[34] Letter from T. H. Morgan, February 10, 1922, Wheeler Papers, HUGFP 87.10, Box 24.

[35] Wheeler, "Present Tendencies in Biological Theory," *Scientific Monthly* 28 (1929): 97–109, at p. 109.

Wheeler continued to describe himself as a field naturalist, even as Jacques Loeb at Chicago came to epitomize the reductionist approach to nature taken by regular visitors to the Woods Hole Marine Biological Laboratory. Although Woods Hole was a field center, it was, as its title suggests, rooted in indoor experimental practice. Some biologists, such as Albert Bethe in Germany, even advocated laboratory studies as the route to a mechanistic understanding of insect psychology. Such approaches were doubly anathema to Wheeler. In metaphysical terms, they did not embrace the phenomena that most interested him: the behavior of live ants in their natural, complex communities. And in epistemological terms, he could not accept a method rooted in atemporal experiment rather than evolutionary narrative.

Wheeler's natural-historical science, then, revolved rather around live organisms and their behavior in the field. Again, it was Forel who had shown the way. The chief practical consequence of Forel's instinct intelligence theory for the student of nature was that it opened up the possibility of evolutionary naturalism. In other words, one could study and compare colonies in varying stages of "primitiveness" in order to learn something about the evolutionary development of mind and behavior. Forel sometimes put this claim extremely strongly; although he did not think that there was a morphological homology between humans and the lower animals, he did claim an analogy proceeding from a "homology of functions" or from adaptation towards an "analogous end." Ants and humans responded to the same pressures of social organization; their societies had to evolve, by hook or by crook, to perform the same tasks necessary to sustain the life of the race. Because behavior and not anatomy provided the key to these analogies, the convergence was to be demonstrated by the naturalist, from living observation and experiment, and not in a laboratory by the morphologist.[36]

Certain ants therefore provided a window on primitive life. As early as 1874, Forel had discussed at length the genera *Ponera, Stenamma, Leptothorax,* and *Myrmecina.* Species of these genera often demonstrated their primitiveness by having no worker caste whatsoever, or, where they did, by having workers rather similar to ordinary females. In addition to this series of intermediary physical forms, Forel pointed out, there existed a parallel series of behavioral intermediaries. He described a graduated series of instincts in the areas of slave making and parasitism, which shadowed the trend of anatomical complexity in the same species.[37] Physiology and behavioral organization evolved in tandem. Species that were undeveloped morphologically speaking, such as

---

[36] Forel, *The Senses of Insects,* p. 1.   [37] Forel, *Les Fourmis de la Suisse,* pp. 440–8.

the Ponerines, allowed one to see correspondingly early stages in the phyletic history of behavior.[38]

Following Forel, Wheeler developed this theme throughout his long career. In *Ants* (1910), Wheeler explained at length the importance of the historical study of evolution.

Although biologists now rarely undertake phylogenetic speculations on a grand scale, they are, perhaps, more active than ever in pursuing such speculations within the more modest confines of species, genera and families. . . . This is particularly true of such compact groups as the Formicidae, which . . . show, both in structure and habits, certain definite, progressive tendencies of development.[39]

Of these subfamilies, the Ponerines had a special significance:

[T]he [subfamily] Ponerinae comprises unmistakably primitive and generalized forms and therefore constitutes a group of two-fold interest, first, as the ancestral stirp of the higher subfamilies, and second, as the oldest existing expression of social life amongst the Formicidae.[40]

Wheeler lamented the fact that the Ponerines were so poorly taxonomized; eventually, in 1931, he formed an expedition to go to Australia, where they were chiefly to be found, and set about filling in these gaps in myrmecological knowledge. In Australia, he wrote, "these ancient insects [the Ponerines] occupy a position among the ants analogous to that of the monotremes and marsupials among mammals. . . . the genus Myrmecia, comprising the "bull-dog ants," . . . may be said to characterize this fauna and at the same time to represent the prototype of all ants."[41] Thus, besides his quasi-Lamarckian mechanism, Wheeler took from Forel the impetus to investigate the mechanisms of evolution through the imaginative reconstruction of the links between primitive and complex organisms. Both men used the analogies provided by extant primitive species to explore the homologous

[38] Forel, *Social World of the Ants*, vol. 1, pp. 1–12, 125–146, and vol. 2, pp. 298–301. Notably, Forel found that Ponerines did not display mutual feeding, his paradigmatic behavior of social life among the ants. See also Bouvier, *Le Communisme chez les Insectes*, pp. 118–19.

[39] Wheeler, *Ants*, p. 225. On pp. 146–7, Wheeler writes: "The distribution of ants may be studied either from a faunistic or from an ethological point of view. . . . Of these two methods . . . the latter leads to more detailed and positive results." See also pp. 156–9.

[40] Ibid., p. 226.

[41] Ibid., p. 227. Wheeler also drew on notes made in New South Wales and Queensland during an earlier trip (1914) in the composition of *Colony-Founding among Ants with an Account of Some Primitive Australian Species* (Cambridge MA: Harvard University Press, 1933).

responses that primitive ancestors of advanced ants had shown in response to the conditions of life. They were natural historians, with the emphasis on history.

IV. THE EUGENIC IMPERATIVE

One might be tempted to imagine that Wheeler's focus on communality had some roots in a left-wing political stance.[42] Nothing could be further from the truth. Wheeler's Lamarckism was constructed in tandem with an elitist view of humanity adapted from the Italian sociologist Vilfredo Pareto, and it complemented his eugenic urges. Though Lamarckism had provided the radical impetus for early transformist thought, for Wheeler, one hundred years later, it was precisely the opposite.[43]

Wheeler constructed his eugenic Lamarckism in a technical sense that was also unlike most earlier Lamarckisms. First, Lamarck was commonly understood to discuss the individual acquisition of traits that could be passed on to the race. This formed the basis for attacks on naïve Lamarckism: cutting the tails of mice did not affect their offspring in the same way. But, as we have already seen in his hierarchies of emergence, Wheeler's theories of ant behavior attempted to account for social phenomena that were irreducible to individual characteristics. Second, Lamarck's theory was generally taken by nineteenth-century radicals to imply an upward push from below. What Wheeler needed was an evolutionary theory about the heritable development of behavior de novo in the social medium. He found this too in Pareto's *Treatise of General Sociology*.[44]

In his treatise, Pareto identified the atavistic, nonintelligent "residues" that got in the way of rational belief formation and action. Residues (the manifestations of "sentiments") were, according to Pareto, emotional responses and non-logical forms of reasoning. A "derivation" was "a non-logical argument, explanation, assertion, appeal to authority, or association of ideas or

---

[42] See Gregg Mitman, *The State of Nature: Ecology, Community, and American Social Thought, 1900–1950* (Chicago and London: University of Chicago Press, 1992) for an account of more intuitively complementary political and natural outlooks on the communal arrangements of social insects.

[43] Adrian Desmond, *The Politics of Evolution* (Chicago and London: University of Chicago Press, 1989).

[44] Wheeler, *The Social Insects*, p. 2. The *Treatise* was published in Italy in 1916, abridged in Italian in 1920, and was made available in English in its entirety in 1935 as *The Mind and Society*, translated by Andrew Bongiorno and Arthur Livingston and published by Harcourt Brace.

sentiments in words."[45] In keeping with the hierarchical evolutionary theories of mind common at the time, instincts were the animal equivalent of residues among humans.[46] In fact, Pareto cited that instinctual paradigm, the social insect, as a prime example of society persisting in the absence of reason.[47] The kinds of urges that prompted an ant to live out its ordered existence were not logical, not consciously foreseen ends. Primitive human societies worked in the same way, and even the majority of "civilized" citizens reasoned no better.

Instinct was the outcome of the unthinking environment in which citizens were raised. Like Wheeler, Pareto put his main emphasis on the behavior of whole systems, rather than on the properties of their component parts. Pareto's initial studies in the mathematics of equilibria were complemented by his extensive reading of Comte, Spencer, and especially of Darwin and Bain. He combined mathematics with his social interests to create an oscillatory model that described economic growth and the distribution of income. The residues and sentiments that he described were the properties of society's members that made their behavior predictable en masse, but it was the large-scale features of society that interested him most.

Wheeler seems to have responded both to Pareto's incorporation of mental atavism and to his holistic attitude, giving both an evolutionary interpretation. Looking about him, Wheeler saw the threat of residues in "the anti-intellectualistic tendencies . . . of European and American thought."[48] He was convinced by Pareto that lack of insight on the part of the common man doomed his society to the degenerate path of fixed instincts, and that he was condemned by his residues to a life that was functionally and psychologically similar to the ant's. Wheeler encouraged the wife of a friend and colleague to translate part of Pareto's book into English, an excerpt then published in the eugenically orientated *Journal of Heredity*. She planned to biologize Pareto's "residues" completely by considering them as hereditary: the atavisms of the masses. It seems likely that Wheeler approved of this gloss.[49]

[45] L. J. Henderson, *On the Social System*, ed. Bernard Barber (Chicago and London: University of Chicago Press, 1970), p. 186.
[46] Vilfredo Pareto, *Compendium of General Sociology*, abridged in Italian from Pareto's *Trattato di Sociologica Generale* (1916) in 1920 by Giulio Farina, English text ed. Elisabeth Abbott (Minneapolis: University of Minnesota Press, 1980), §713.
[47] Pareto, *Compendium of General Sociology*, §§63–90.
[48] Wheeler, *The Social Insects*, p. 2.
[49] This woman was Marion Fairchild, the wife of David Fairchild. David Fairchild worked with Wheeler for a short time on termites and, if his letters are to be believed, always wished that he had done more of this. Instead, he went on to become a vocal conservationist and eugenist, but he retained a close friendship with Wheeler. They shared many social convictions, and Fairchild always appreciated the philosophy to which Wheeler introduced him. This

In sociological terms, evolution provided reasons to suppose that human behavior had hardened into forms that worked. In other words, residues were equivalent to the instincts that enabled a society – ant or human – to cohere. This was precisely what Forel had said about instincts, though Paretan instincts, of course, carried a strongly negative value judgment: they were the sign of the intellectually lazy, if not the outright incapable. A society defined its subsequent caste arrangements, so it was likely to continue in its residue-infested state, open to manipulation from above.

When Pareto's sociology was added to his earlier philosophical brew, Wheeler arrived at a biological theory explaining why society emerged, evolved, and then determined the lives and instincts of its "individual" members. This message was the undercurrent to *Foibles of Insects and Men* (1928), and to his unfinished manuscript about human gullibility, *Holy Bluff*. Late in life, Wheeler even turned to psychoanalysis in an attempt to find a higher-level discipline that dealt with the residues within humanity, disrupting man's claims to behave with individual intentionality.[50]

The upshot of Pareto's work was that residues ought not to be dismissed; rather, they required elucidation by sociologists, for they held the key to the understanding and effective governance of society. Pareto did not appeal to the average man to uncover his residues but spoke instead to those who would lead or comment upon society. Wheeler's representation of nature was similarly constructed in conjunction with his self-representation as an expert natural historian. This connection was explicitly acknowledged – even intended – by Wheeler himself. He wrote:

During the nineteenth century biology and sociology developed in rather intimate symbiosis. Though Comte founded sociology on biology, it is well known that certain important conceptions, such as the struggle for existence, the survival of the fittest and the physiological division of labor, were derived from sociological sources and later extended to the entire world of organisms in the Darwinian theory of evolution. If we may judge from the works of Spencer, Espinas, de Lilienfeld, De [sic] Greef, Worms,

ideological and philosophical common ground puts Wheeler's contribution to the question of the individual and the mass in a eugenic context. Wheeler's correspondence with Fairchild is held at the Wheeler archive in the Pusey Library, Harvard. For an introduction to eugenic issues during the period, see Daniel J. Kevles, *In the Name of Eugenics: Genetics and the Uses of Human Heredity* (New York: Knopf, 1985); and Diane B. Paul, *Controlling Human Heredity, 1865 to the Present* (Atlantic Highlands, NJ: Humanities Press, 1995).

50 Meanwhile, Freud and Ferenczi contemplated a Lamarckian explanation that placed human telepathy midway between the ants' mental unity and the consulting room phenomenon of transference. See Pamela Thurschwell, "Ferenczi's Dangerous Proximities: Telepathy, Psychosis and the Real Event," *differences* 11 (1999): 150–78.

Waxweiler and others, this theory, after its first clear enunciation, seems to have been more heartily welcomed and embraced by the sociologists than by the biologists. . . .[51]

Wheeler's project, then, was to reunite the two fields. In doing so, Wheeler gave himself the authority to comment upon human affairs because of his biological expertise.[52] The true "top-down" method of sociology places the investigator outside the society that he observes – a situation that is also referred to as scientifically objective. But what exactly did Wheeler want to say from his expert perspective? His focus on cooperation did not lead to him to socialism, as it had some Victorians, with their emphasis on "mutual aid." In this specific historical context, the new or rediscovered relationship between biology and sociology brought with it a eugenic imperative and a dismissive attitude towards the "masses."[53]

Eugenic arrangements were a part of the functional analysis of society, carried out by those in the elite caste, whose members and their intellectual attributes had been unconsciously created by the whole of society during its long history:

It is probably not a mere coincidence that we should be most diligently discussing eugenics, or the restriction of reproduction to the sane in mind and body, at a time when we are most exercised by the high cost of living. Did space permit, it could be shown that man, like other social organisms, has for ages sought and is still seeking means of regulating the reproductivity of his race to prevent its exceeding its food supply . . . [such as] monasticism, wars . . . [and] religious, property and caste restrictions to marriage[,] [all of which] have been only partially successful.[54]

In 1919, Wheeler gave a strange lecture in which he played with the notion of such natural eugenic control. His paper, "The Termitodoxa, or Biology and Society," took the form of a termite narrating the history of his race.[55] The race was degenerating terribly until the biologists told everyone else how they should be organized – thus producing the extant successful caste arrangements of termites. Three years later, Wheeler took on this role in real life, accepting a

[51] William M. Wheeler, "Animal Societies: Biology and Society," *Scientific Monthly* 39 (1934): 289–301, at p. 290.
[52] Compare this to role of social expertise sought by other biologists of the time. See Gary Werskey, *The Visible College: A Collective Biography of British Scientists and Socialists of the 1930s* (London: Free Association, 1988).
[53] For a coy account of Wheeler and the eugenics movement, see Mary A. Evans and Howard E. Evans, *William Morton Wheeler, Biologist* (Cambridge, MA: Harvard University Press, 1970), pp. 243–52.
[54] William M. Wheeler, "Notes about Ants and Their Resemblance to Man," *National Geographic Magazine* 23 (1912): 731–66, at pp. 742–3.
[55] Wheeler, "The Termitodoxa."

position on the advisory council of the Eugenics Society of the United States of America at the time of its foundation. The literature sent by the committee to Wheeler was mainly a polemic against the immigration of "inferior" races. Wheeler wrote, "It is a very great honor [to serve on the advisory council] and I shall be glad to do what I can in aiding the work of the Committee."[56] Other members of the committee included Wheeler's close friend (and former entomological colleague) David Fairchild and his close friend in latter years the evolutionist, philosopher, and sometime entomologist George H. Parker. The entomologists Vernon Kellogg and Anne H. Morgan each took a place on the committee, as did at least two other acquaintances of Wheeler's – Robert Yerkes and William McDougall. Interestingly, Charles Davenport, a prominent eugenist and member of the advisory council, was registered as a member of the Entomological Society in 1910.[57] In 1933, Wheeler gave a paper on "Animal Societies" at the Biology and Society symposium organised by the American Society of Naturalists. While Wheeler's paper was an abstract discussion of the issues that all societies needed to solve (notably, the "problem of the male"), the other papers at this conference were overt discussions of applied eugenics.[58] They considered which races were inferior and which components of American society were of similarly low hereditary caliber. Thus Wheeler contributed to debate about human society. He also contributed a chapter to a book on *Human Biology and Racial Welfare*.[59] Additionally, Wheeler's personal correspondence reveals his engagement with eugenic topics.

Wheeler had not wanted to identify himself as a member of the masses whose behavior he had discovered through his work with ants. Instead, his role, like the termite biologists' of the "Termitodoxa," was to direct society – to take advantage of the mass manipulability of its unthinking members for their own long-term benefit. The objective scientist was therefore not like most people by his very nature. Unlike Forel, Wheeler had to place himself outside the human formicary *because of his thoroughgoing sociological approach*; indeed, he argued on several occasions that the scientist was innately antisocial. Here we see Wheeler's particular development of a group-based analysis of social evolution, moving beyond "orthodox" Lamarckism to applied elitist sociological theory.

---

[56] Letter from William M. Wheeler to Irving Fisher, January 19, 1924, Wheeler Papers, HUGFP 87.10, Box 12.

[57] Ibid.

[58] The three papers were printed in *The Scientific Monthly* 39 (1934): 289–322.

[59] William M. Wheeler, "Societal Evolution," in E. V. Cowdry, ed., *Human Biology and Racial Welfare* (New York: Hoeber, 1930), pp. 139–55.

V. CONCLUSION

Wheeler's *Lamarck Manuscripts* are introduced by a hilariously venomous biography of Darwin, written by F. G. Crookshank (like Wheeler, a colleague of their publisher C. K. Ogden). Crookshank claimed that Darwin's whole theory of an arbitrarily judgmental nature was merely a guilty projection of his own hypochondriacal existence:

Darwin's philosophy, which ascribes all to the play of external forces or innate characters, is only the exteriorization of his neurotic excuses for his own life, carried on successfully . . . in a fashion only possible for a man in easy circumstances, who had taken the precaution to surround himself with the self-sacrificing devotion of an adoring wife. No man could live happily the life that Darwin lived did he not feel that God, or Nature, like his father, had been a "little unjust" to him. . . . His greatest fiction of all, save the *Origin of Species*, was that his mysterious illness was a cross laid on him by his "constitution."

By comparison, Lamarck's biography was a tale of courageous triumph over disadvantage. Why was Wheeler pleased to have Darwin denigrated so? The contingencies of history are largely to blame. Darwin had been appropriated by laboratory biologists, who now threatened to discredit Wheeler's entire approach to the study of nature.[60]

One might have expected Wheeler to embrace Darwin, a natural historian in both his metaphysical and epistemological senses, as an appropriate intellectual forebear. Instead, he ridiculed Darwin's character and downplayed his achievements. Though Wheeler considered that Darwin's work had saved natural history from "scholasticism," he claimed that Darwin's description of evolution could easily have been given by other members of the scientific community had they paid more attention to the physical and chemical phenomena of life.[61] Wheeler's fundamental problem was that categories of biology had been defined for him by his opponents, his competitors for scientific prestige. He knew that he was not a laboratory biologist; he knew that many laboratory biologists embraced Weismannism; thus Weismannism, or neo-Darwinism, came to stand for many things from which Wheeler wished to distance himself. Wheeler's "eighth mortal sin" confirms the natural-historical nature of the ninth. The eighth unforgivable thing in biology, joked Wheeler in 1917,

---

[60] Perhaps there was also a hint of the rejection of British authority here. Darwin's life in a Kent country house ill fitted the Rooseveltian tradition of manly natural history prevalent in the United States.

[61] William M. Wheeler, "Predarwinian and Postdarwinian Biology," *Popular Science Monthly* 74 (1909): 381–5.

was anthropomorphism – a charge frequently leveled at "mere" naturalists.[62]
Wheeler's two gibes at the dogmas of the church of biology – and he *loathed*
organized religion – suggest strongly that the two are linked. Knowing that
one, anthropomorphism, was considered the besetting sin of the naturalists,
it seems fair to conclude that Wheeler associated the other, Lamarckism, with
the same group of people.

---

[62] William M. Wheeler, "A Study of Some Young Ant Larvae with a Consideration of the Origin
and Meaning of Social Habits among Insects," *Proceedings of the American Philosophical Society*
57 (1918): 293–343, at p. 293. See also Ralph Lutts, *The Nature-Fakers: Wildlife, Science and
Sentiment* (Charlottesville and London: University Press of Virginia, 1990).

CHAPTER NINE

Contemporary Darwinism and Religion

Mikael Stenmark

The relationship between Darwin's theory of evolution and religion has been, to say the least, a controversial topic ever since the publication of *On the Origin of Species* in 1859. Interestingly enough, evolutionary biologists have had and continue to have quite different views about this relationship. The questions that I want to address in this paper are: (1) what views about the proper relationship between science and religion can we find among contemporary evolutionary biologists? and (2) how should we assess these views? – more specifically, which one (if any) is the most reasonable one to adopt? In relation to these issues, I shall also ask (3) what would count as Darwinian heresy on this matter?

Two radically different perspectives on these issues can be found among evolutionary biologists. On the one hand, we have Darwinians, such as Stephen Jay Gould, who hold that religion and evolutionary biology (or, more broadly speaking, science) are logically distinct and fully separate domains with different subject matters, methods, and aims. On the other, we have those such as Edward O. Wilson and Richard Dawkins, who think that science in general, and especially biology, severely undermines traditional religion and that science, to some extent, can even replace religion. Let us first look at these views in more detail and then assess them critically. Let us also ask whether there is any other way of understanding the relationship between science (and, in particular, evolutionary biology) and religion.

In writing this chapter I gratefully acknowledge the financial support of the Swedish Research Council.

I. DARWINIAN RESTRICTIONISTS

In *Rocks of Ages: Science and Religion in the Fullness of Life* (1999), Stephen Jay Gould has delivered one of the most recent statements about how we ought to understand the relationship between contemporary science and religion. He thinks that the idea that there has been, and still is, a war going on between science and religion is wrong. It fails both as a historical account of how science and religion have been related and as a normative account of how science and religion ought to be related. Instead, he maintains that each inquiry frames its own questions and criteria of assessment. Gould writes that

the net, or magisterium, of science covers the empirical realm: what is the universe made of (fact) and why does it work this way (theory). The magisterium of religion extends over questions of ultimate meaning and moral value. These two magisteria do not overlap, nor do they encompass all inquiry. . . . [1]

Science and religion ask different kinds of questions, and to that extent they have different aims and subject matters.

Science and religion also have methodologies or epistemologies. Both scientists and religious people have to regulate what they believe in some way. They use standards of assessment of some kind; let us call these standards "epistemic norms." Gould maintains that science and religion have different epistemic norms and that we have to acknowledge and accept this difference, without imposing one magisterium's norms on the other magisterium.

Gould contrasts the way in which the disciple Thomas's request for evidence is evaluated in Christian practice with how a similar request is evaluated in scientific practice. When the other disciples told Thomas that they had met the resurrected Jesus, Thomas responded by saying, "Unless I see the nail marks in his hands and put my finger where the nails were, and put my hand into his side, I will not believe" (John 20:25). A week later Jesus reappeared, and this time Thomas was also present. Jesus let Thomas put his finger where the nails had been and put his hand into his side, and then Thomas believed and said to Jesus, "My Lord and my God." Jesus responded by saying, "Because you have seen me, you have believed: blessed are those who have not seen and yet have believed" (John 20:29). Gould accepts this as a proper epistemic norm of religion but also writes that he "cannot think of a statement more foreign to the norms of science. . . . A skeptical attitude toward appeals based only on authority, combined with a demand for direct evidence . . . represents the first commandment of proper scientific procedure."[2]

---

[1] Stephen Jay Gould, *Rocks of Ages: Science and Religion in the Fullness of Life* (New York: Ballantine, 1999), p. 6

[2] Ibid., p. 16.

Despite this difference, Gould maintains that both science and religion are important and necessary if we want to reach a full understanding of human life in all its complexity. Gould, moreover, suggests that we should accept what he calls *the principle of NOMA* (or *Non-Overlapping Magisteria*).[3] According to the principle of NOMA, the relationship between science and religion ought to be one of respectful noninterference. The principle could, roughly, be explicated as follows:

> *Both science and religion are valid human inquiries and ought to be respected but treated as logically distinct and fully separate areas of inquiry with their own questions and epistemologies (or methodologies).*

This principle puts certain restrictions on what kinds of claims religious believers or scientists can make *as* religious believers or scientists. That is to say, there can be both misuse of science and misuse of religion. Religion is misused when it is used as a *control belief* in scientific inquiry, that is, as a way of restricting the kind of factual conclusions that scientists are allowed to draw from the data they have access to. This is done, for instance, when religious believers reject the theory of evolution because it does not fit with their understanding of what the Bible teaches and therefore they want to impose on science a different research program ("creationism") that better fits their religious convictions.

But Gould also thinks, perhaps a bit more surprisingly, that religion is misused – or, better, that NOMA is violated – when religious believers adhere to a certain conception of God. This is so because he maintains that the first commandment of NOMA is: "Thou shalt not mix the magisteria by claiming that God directly ordains important events in the history of nature by special interference knowable only through revelation and not accessible to science."[4] Thus religious believers cannot properly claim that God's action sometimes results in the occurrence of a miracle. Moreover, "people whose concept of God demands a loving deity, personally concerned with the lives of all his creatures" also violate NOMA, although in "a more subtle" way.[5] This means, I think, that religious people should not understand God's personal concern for them and others in such a way that they believe that God has prearranged natural history – for instance, the origin of the human species, or the birth or death of particular individuals – so that it will have a certain outcome.

Gould maintains that science can be misused as well. Scientists can in their profession violate the principle of NOMA. What NOMA does is to "forbid

---

[3] Gould's definition of magisterium is: "A magisterium . . . is a domain where one form of teaching holds the appropriate tools for meaningful discourse and resolution" (ibid., p. 5).
[4] Ibid., p. 84.   [5] Ibid., p. 93.

scientific entry into fields where many arrogant scientists love to walk, and yearn to control."[6] Gould thinks, in fact, that many contemporary biologists have imperialistic aims. They are what we might call *Darwinian expansionists*, because they attempt to expand the boundaries of evolutionary biology in such a way that it covers other areas of inquiry – for instance, ethics and religion.[7] One example that Gould comes back to several times is the attempt by some Darwinians to provide answers to moral questions. But this is to misuse evolutionary biology: "Any argument that facts or theories of biological evolution can enjoin or validate any moral behavior represents a severe misuse of Darwin's great insight, and a cardinal violation of NOMA."[8] Gould believes that the same is often true with respect to religion. He confesses that he is "discouraged when some of [his] colleagues tout their private atheism . . . as a panacea for human progress against an absurd caricature of 'religion,' erected as a straw man for rhetorical purposes."[9]

So it seems – even if Gould omits to state it – that the second commandment of NOMA is: "Thou shalt not mix the magisteria by claiming that science directly ordains solutions to moral and existential concerns by special interference knowable only through scientific experiments and discoveries that are not accessible to religion."

Gould thinks instead that Darwin should serve as the model for biologists (as well as for scientists in general):

Darwin did not use evolution to promote atheism, or to maintain that no concept of God could ever be squared with the structure of nature. Rather, he argued that nature's factuality, as read within the magisterium of science, could not resolve, or even specify, the existence or character of God, the ultimate meaning of life, the proper foundations of morality, or any other question within the different magisterium of religion.[10]

We can, therefore, identify both religious heresy and scientific heresy on this issue. One becomes a heretic when one violates the principle of NOMA by trying to expand one's own magisterium into the other's magisterium. Thus a *Darwinian heretic* would be a biologist who, in the name of science, uses evolution to promote atheism, theism, or any other solution to our existential concerns, or who uses evolution to specify the proper foundation or content of morality or to reject any moral discourse whatsoever.

---

[6] Ibid.

[7] The term "expansionists" comes originally from Loren R. Graham, *Between Science and Values* (New York: Columbia University Press, 1981), p. 6.

[8] Gould, *Rocks of Ages*, p. 163.   [9] Ibid., p. 209.

[10] Ibid., p. 192.

II. DARWINIAN EXPANSIONISTS

The idea that evolutionary theory has great implications for human society and our self-knowledge has, however, always been a part of the Darwinian tradition. Contemporary biology is no exception. Thus, Richard D. Alexander talks about the recent developments within evolutionary biology as the "greatest intellectual advance of the twentieth century" that should have a profound impact on our self-view and our understanding of morality.[11] In fact, he believes that we have to "start all over again to describe and understand ourselves" and that we have to do it "in terms alien to our intuitions."[12] Richard Dawkins writes that because we have evolutionary theory, "We no longer have to resort to superstition when faced with the deep problems: Is there a meaning to life? What are we for? What is man?"[13] Moreover, he agrees with the eminent zoologist G. G. Simpson that "all attempts to answer that question ["What is man?"] before 1859 are worthless and that we will be better off if we ignore them completely."[14]

What is it that evolutionary biology can teach us that goes beyond the empirical questions that Gould thinks it should be occupied with? There are several things. Evolutionary theory is taken to be able to show that morality is ultimately about selfishness or maximizing fitness. Michael Ruse and Edward O. Wilson tell us that evolutionary biologists have discovered that "in an important sense . . . ethics is an illusion fobbed off on us by our genes to get us to cooperate" and that therefore there is no objectivity to morality.[15] We are deceived by our genes into thinking that there is a disinterested, objective, and binding morality, which we all should obey.[16] Dawkins agrees and tells us that evolutionary theory supports the idea that life is selfish all the way down; thus, we can even talk about selfish genes. Moreover, since we are "survival machines" that are "blindly programmed to preserve [these selfish] genes," no matter how much "we wish to believe otherwise, universal love and the welfare of the species as a whole are concepts that simply do not make

---

[11] Richard D. Alexander, *The Biology of Moral Systems* (New York: Aldine De Gruyter, 1987), p. 2.

[12] Ibid., p. 2.

[13] Richard Dawkins, *The Selfish Gene*, 2nd ed. (Oxford: Oxford University Press, 1989 [1976]), p. 1.

[14] Ibid.

[15] Michael Ruse and Edward O. Wilson, "The Evolution of Ethics," in James E. Huchingson, ed., *Religion and the Natural Sciences: The Range of Engagement*, (Forth Worth, TX: Harcourt Brace, 1993), p. 310.

[16] Michael Ruse and Edward O. Wilson, "Moral Philosophy as Applied Science," *Philosophy* 61(1986): 179.

evolutionary sense."[17] No wonder Alexander thinks that these claims, if true, would radically change our self-view.

Dawkins also proclaims that Darwinism makes it possible to be an "intellectually fulfilled atheist" and that because evolutionary theory undermines, if not refutes, traditional religious beliefs by showing that the universe lacks design or purpose, biologists ought to be atheists.[18] Although Wilson thinks that religion constitutes the greatest challenge to biology, he maintains that we can "explain traditional religion . . . as a wholly material phenomenon" by using evolutionary theory.[19]

Not only can Darwinism, Wilson tells us, be used to explain religion as a strategy solely adapted to secure genetic fitness, it can even replace traditional religion. The evolutionary epic provides us with a new mythology, and it can constitute the key element in our new religion – what Wilson sometimes calls "scientific materialism" and at other times "scientific naturalism." But Wilson does not think that it is possible now to predict the forms that religious life and rituals will take as "scientific materialism appropriates the mythopoeic energies to its own ends."[20]

It is thus not merely the case that Gould fails to realize the full potential of Darwinism; he also fails to understand what religion is, or at least what kind of religion it is that ought to be taken seriously. Dawkins tells us that he pays

religions the compliment of regarding them as scientific theories. . . . I see God as a competing explanation for facts about the universe and life. This is certainly how God has been seen by most theologians of past centuries and by most ordinary religious people today. . . . Either admit that God is a scientific hypothesis and let him submit to the same judgement as any other scientific hypothesis. Or admit that his status is no higher than that of fairies and river sprites.[21]

Moreover, Dawkins thinks that scientists can use scientific methodology to criticize religious attitudes and epistemic norms. He tells us that faith

means blind trust, in the absence of evidence, even in the teeth of evidence. The story of Doubting Thomas is told, not so that we shall admire Thomas, but so that we can admire the other apostles in comparison. Thomas demanded evidence. . . . The other

---

[17] Dawkins, *The Selfish Gene*, pp. v, 2.
[18] Richard Dawkins, *The Blind Watchmaker* (New York: Norton, 1986), pp. 5–6.
[19] Edward O. Wilson, *On Human Nature* (Cambridge, MA: Harvard University Press, 1978), p. 192.
[20] Ibid., p. 206.
[21] Richard Dawkins, "A Reply to Poole," *Science and Christian Belief* 7(1995): 46–7.

apostles, whose faith was so strong that they did not need evidence, are held up to us as worthy of imitation.... Blind faith can justify anything.[22]

Presumably, this means that if religious believers were really rational – and on this point they can learn a lot from scientists – then they would admire Thomas and not the other disciples and consequently change their epistemic norms in such a way that they would resemble scientific norms.

According to these biologists, Gould has got most things wrong about the proper relationship between science and religion. Science and religion are, contrary to what Gould thinks, on the same turf. Two strategies are used (either separately or jointly) to show that there is a union of domains. Either it is argued that traditional religion offers rival explanations about empirical phenomena, or it is maintained that the boundaries of contemporary science (especially of evolutionary biology) can be expanded in such a way that it covers or will eventually cover not only empirical questions but also moral and existential questions.[23]

III. THREE SCIENCE/RELIGION VIEWS

What should one think about these radically different claims about the boundaries of contemporary science – especially of evolutionary biology – and its proper relationship to religion? It seems that we can choose between three options: (1) there is no overlap between science and religion (including morality); (2) there is, more or less, a union of the domains of science and religion (including morality); or (3) there is some overlap between the domains. Call the first the *independence view*, the second the *conflict view*, and the third the *contact view*.

Which view is the most reasonable one? In order to be able to answer this question, we have to know in more detail what kinds of activities science and religion are. Whatever else they might be, science and religion are complex activities performed by human beings in cooperation within a particular

[22] Dawkins, *The Selfish Gene*, p. 198.

[23] Related to this point, it is important to notice that even if all evolutionary biologists discussed in this section are Darwinian expansionists, in contrast to Gould, they do not agree in what way, exactly, the boundaries of evolutionary theory ought to be extended. Whereas, for instance, Ruse and Wilson think that evolutionary theory can be extended to offer solutions to our moral questions, this is something that Alexander and Dawkins deny (Alexander, *The Biology of Moral Systems*, p. xvi; Dawkins, *The Selfish Gene*, p. 2). Moreover, whereas Dawkins and Wilson think that evolutionary theory can be extended to offer solutions to our existential questions and to refute traditional religion, this is something that Ruse explicitly denies (Michael Ruse, *Taking Darwin Seriously*, 2nd ed. [Oxford: Blackwell, 1998 (1986)], p. 294).

historical and cultural setting; in short, they are social practices. As social practices, they are performed by certain groups of people. These groups of people (the practitioners) are organized in particular ways. The practices can be defined by identifying the goals that the practitioners have more or less in common and by the means that they develop and use in order to achieve these goals. Further, these practices have histories, and they therefore constitute traditions. Thus one possible level of intersection is *social*. We can find possible overlaps when it comes to who participates in these practices and how these practices are socially structured: what functions different groups of practitioners play, how knowledge or something else essential for the practice is transmitted from one generation to the next, and so on.

In what way is this relevant to our discussion? Gould would hardly deny that there is a social intersection in the sense that the same people can participate in both religious and scientific practices. This could not be a part of his claim that these two magisteria do not overlap. Nevertheless, it is relevant, because Gould points to the actual *social* intersection as a reason for thinking that there is no *methodological* and *theoretical* intersection or, more exactly, that there is no warfare between science and religion in terms of either rationality or areas of inquiry. He asks rhetorically, "if science and religion have been destined to fight for the same disputed territory" how could it then be possible that "science, at the dawn of the modern age, [has been] honorably practiced by professional clergymen (who, by conventional [warfare] views, should have undermined rather than promulgated such an enterprise)?"[24] His argument seems to be:

1. If the warfare view (or the conflict view) is true, one would expect that people who were deeply religious would not be in the forefront of developing science.
2. But people who were deeply religious – such as Galileo, Newton, Faraday, and Eddington – were in the forefront of developing science.
3. Therefore, we ought to reject the warfare view and accept an independence view based on the principle of NOMA.

This argument, I think, undermines the plausibility of a view that says that there has always been warfare between science and religion. But it is not quite as strong as Gould seems to believe. For one thing, a defender of the warfare view could argue that the recent developments in, for instance, evolutionary biology are such that there is *now* a genuine conflict between science and

---

[24] Gould, *Rocks of Ages*, p. 70.

religion. Thus, William Provine thinks it is true that "very few truly religious evolutionary biologists remain. Most are atheists, and many have been driven there by their understanding of the evolutionary process and other sciences."[25]

In other words, *science (and, of course, religion as well) is an activity that changes over time, and therefore, whether or not there is a conflict between science and religion depends, in particular, on the specific content of the scientific theories (and, of course, also on the specific content of the religious beliefs) accepted at a given time.* Consequently, Gould's argument is not sufficient to establish the conclusion that the proper relationship between science and religion ought to be guided by the principle of NOMA. There is no shortcut possible on this issue. If he wants to convince us that NOMA also applies to the relationship between contemporary evolutionary biology and religion, then he needs to respond to the claims of his expansionistic colleagues. Thus, what might have counted as Darwinian heresy in Darwin's days might not count as heresy today because of the recent development of biology. (By "heresy" I mean in this context an illegitimate expansion of evolutionary theory into new areas of inquiry, such as morality and religion.)

At least two conclusions follow. First, the ways in which religion and science are related will vary during the course of history, because both are dynamic and evolving social practices and any plausible normative account must take that into consideration. Second, the social overlap between science and religion undermines any idea that we will always find scientists and religious believers in simple opposition.

But one could still maintain, as Provine does, that the number of religious evolutionary biologists is shrinking. If this is true, the interesting philosophical question is, of course, whether the number is shrinking for good reasons, or, more exactly, whether it is shrinking because contemporary evolutionary biology directly implies the refutation of religion. That is, is the number of religious evolutionary biologists shrinking, as Provine thinks, for good *scientific* reasons? Let us initiate an inquiry that might eventually lead to an answer to this question by focusing on the issues, raised by Gould, about the proper epistemic norms and areas of inquiry of science and religion.

IV. THE EPISTEMIC NORMS OF SCIENCE AND RELIGION

Beliefs, theories, and the like are acquired, revised, or rejected in the actual life of both science and religion. These processes involve reasoning of some

---

[25] William Provine, "Evolution and the Foundation of Ethics," *MBL Science* 3 (1988): 28.

sort. Do practitioners in the two fields endorse the same kinds of reasoning, or, if not, should they endorse the same kinds of reasoning? In particular, is it legitimate for scientists to critically challenge the way in which beliefs are formed, rejected, and revised in religion by taking science as the paradigmatic example of rationality?

Gould, as we have seen, denies this by maintaining that the epistemic norms of one magisterium should not be imposed on the other magisterium; the relationship ought to be one of respectful noninterference. He acknowledges that the norms of religion are foreign to the norms of science, but he refuses to pass any judgment on those norms. Other Darwinians, on the other hand – such as Dawkins and Wilson – feel no such obligation and quite straightforwardly maintain the superiority of the epistemic norms of science.

But Dawkins, Gould, and Wilson seem to agree that rational scientists endorse what philosophers call *evidentialism* as their model of rationality. Evidentialism is, roughly, the view that it is rational to accept a theory or belief only if, and to the extent that, there are good reasons (or evidence) to think that it is true.[26] Gould believes evidentialism to be mandatory in science, but he thinks that it is, at best, permissible within religion, because the religious ideal is the one embodied not in Thomas's but in the other disciples' response to the resurrected Christ. The message that the Biblical narrative delivers, according to both Gould and Dawkins, is that the proper way to acquire a belief within the magisterium of religion is to accept it on the basis of trust and authority, and not on the basis of evidence. Dawkins is very critical of such an epistemic norm; he argues that this kind of faith, this "blind trust" that he thinks the other apostles express (their faith was so strong that they did not need evidence), is not worthy of imitation, because it can justify anything.[27]

Suppose for a moment that Dawkins is right that religious faith is a matter of blind trust. Why would it not be appropriate for him to suggest that a different epistemic norm, one that has proved to be successful within science, would be a better norm to adopt within religion as well? If we can improve our cognitive performance in this way – by taking what we have learned in one area and applying it in another area of life – it is hard to understand why anyone would object. Whether such an attempt to impose the epistemic norms or methods of one practice on another practice will prove to be convincing,

---

[26] See Mikael Stenmark, *Rationality in Science, Religion, and Everyday Life: A Critical Evaluation of Four Models of Rationality* (Notre Dame, IN: University of Notre Dame Press, 1995), Chapters 3–7.

[27] Dawkins, *The Selfish Gene*, p. 198.

however, depends on *whether one has sufficiently understood what is going on in this other practice.* We sometimes respond very negatively to a proposed "improvement" of a particular practice precisely because we anticipate a lack of awareness and sophistication in understanding what the context and the objectives of the practice in question are.

This applies to academic disciplines as well. Wilson, for instance, argues that the explanatory categories and methods of evolutionary biology ought to be extended into the social and human sciences. He writes, "It may not be too much to say that sociology and the other social sciences, as well as the humanities, are the last branches of biology waiting to be included in the Modern Synthesis."[28] If this is done properly, then, presumably, the cognitive performance of the scholars working in these fields can be improved. It is unclear to what extent Wilson wants to replace, for instance, the traditional methods of sociology with biological methods. But suppose this meant that sociologists could keep their statistical and mathematical methods but had to replace their hermeneutic methods with biological methods. Under such conditions, it is fully understandable that many sociologists would object to the attempt to impose biological methods on sociology, by maintaining that Wilson fails to do justice to the subject matter of sociology. Some of the data sociologists try to understand consist of meaningful phenomena, such as texts. These texts include legal documents, letters, political manifestos, and so on. Thus, it is reasonable to believe that some of the phenomena studied by sociologists are not detectable by biological means; they escape biological methods. Thus, hermeneutic methods cannot be replaced by biological methods.

The same kinds of considerations are, of course, relevant in the case of religion. Have, then, Dawkins and Gould properly understood the epistemic norms of religion? Let us consider in more detail the biblical narrative that Gould and Dawkins believe illustrates a central religious epistemic norm – namely, the story of doubting Thomas. Dawkins takes the difference to be that Thomas (as scientists do) demanded evidence, whereas the other disciples had a faith so strong that they did not need evidence, and that therefore their faith expressed blind trust.[29] Gould takes the difference to be that Thomas (as scientists do) demanded evidence and had a skeptical attitude toward appeals based only on authority, whereas he (like the other disciples) should have known by faith that Jesus was resurrected and alive.[30] Thus, the epistemic norm of religion in question is taken to be, roughly, that *a religious belief ought*

---

[28] Edward O. Wilson, *Sociobiology* (Cambridge, MA: Harvard University Press, 1975), p. 4.
[29] Dawkins, *The Selfish Gene*, p. 198.   [30] Gould, *Rocks of Ages*, p. 16.

*to be accepted only (in Dawkins's case) or preferably (in Gould's case) on the basis
of trust and authority, and not on the basis of evidence.*

But is it really reasonable to believe that the conclusion we should draw
from this biblical narrative is (in Dawkins's case) that the other disciples had
a faith that was so strong that they did not need evidence, or (in Gould's
case) that their faith in Christ was based only on authority? I do not think
so, because if we return to the text, we can actually read about how the other
disciples arrived at their faith in the resurrected Christ. We can read that when
they were gathered in a room, "Jesus came and stood among them and said,
'Peace be with you!' After he said this, he showed them his hands and side.
The disciples were overjoyed when they saw the Lord" (John 20:19–20). But
if this is true, then they did not violate the evidentialist norm. They held their
belief on the basis of evidence and not merely on the basis of authority.

What, then, is problematic from a religious point of view about Thomas's
doubt, if it is not that he attempted to base his faith on evidence whereas the
other disciples had no such intention? It seems to be two things.

First, Thomas questioned the testimony of his close friends, the other
disciples, with whom he had long lived and suffered hardship. When the
other disciples told Thomas that they had met the resurrected Jesus, he did
not believe them and asked for more evidence, in this way indicating, in a
situation that critically tested their friendship, that he did not really trust
them.

Second, his skepticism was too severe; his demand for evidence was beyond
what one should require in order to be convinced of something, a theme
well captured in Mark Tansey's painting *Doubting Thomas*, which Gould also
draws to our attention.[31] In 1986, Tansey depicted a man who refuses accept
the theory of continental drift in general, or even the reality of earthquakes
in particular. An earthquake has fractured both a road and the adjoining cliff,
but the man still doubts. So he instructs his wife to straddle the fault line with
their car, while he gets out and thrusts his hand into the crack in the road.
Then, but only then, he believes.

How skeptical one should be in life is a difficult question to settle. The
price of skepticism is the risk of failing to hold a substantial number of true
beliefs, whereas the price of credulity is the risk of ending up with too many
false beliefs. But contrary to both Dawkins and Gould, I think that neither the
biblical version nor Tansey's version of the doubting Thomas story provides
a very good model even for evolutionary biologists. Many theories accepted

---

[31] Ibid., p. 14f.

in evolutionary biology are such that the evidence they are based on, to put it crudely, is not such that one can feel it directly with one's hands, and the conclusions (i.e., the theories) always go beyond the actual evidence. Nor would they typically, I believe, be worthy of imitation in everyday situations. The reason for this is that we have finite cognitive resources and limited time at our disposal, but we still need to believe and do a number of everyday things in order to function properly.[32] We would not have time to do most of these things if the norms of rationality were set on the level of either the ancient or the modern doubting Thomas. Moreover, there are people in our Western society who do not believe that the Holocaust happened, or that Julius Caesar was a Roman emperor. Why they do not believe these things may be hard to know, but it could be that they do not think there is sufficient evidence to support these beliefs. If so, we would consider such people irrational, not because of their credulity, but because of their extreme skepticism.

There are good reasons to question that the conclusion we should draw from the biblical narrative is (in Dawkins's case) that the other disciples had a faith that was so strong that they did not need evidence or (in Gould's case) that their faith in Christ was based only on authority. Nevertheless, I think – to return to the first theme, what makes Thomas's doubt problematic from a religious point of view – Dawkins and Gould have captured one feature of religious belief regulation that makes it different from scientific theory regulation. There are, however, reasons to believe that Dawkins's negative assessment of this feature of religious belief regulation is somewhat premature. The difference is this: the critical questioning of the beliefs of other people (exemplified by Thomas) is typically regarded as an epistemic virtue within science, whereas this is not normally the case in religion. Gould maintains, as we have seen, that the first commandment of science is a demand for evidence and a skeptical attitude toward appeals based only on authority, and Dawkins thinks that Thomas's demand for evidence is what is worthy of imitation and that this is what characterizes the scientific attitude. In religion, on the other hand, people frequently believe things on the basis of authority, discourage critical questioning, and talk instead about trust.

Should we then adopt the skeptical scientific norm in religion, and perhaps also in every other practice we participate in? Not necessarily. In fact, this might even turn out to be quite imprudent. In order to see this, compare two scenarios. In the first scenario, imagine that one of Dawkins's colleagues in zoology tells him that she has conceived a great new theory according to which

---

[32] See Stenmark, *Rationality in Science, Religion, and Everyday Life*, Chapter 8, for a discussion how this affect the formulation of an appropriate model of rationality.

genes cause certain animals – say, chimpanzees – to behave in a particular way. This theory, however, runs contrary to most things zoologists previously have thought. Dawkins says, "Great, very interesting indeed. But what evidence do you have that supports your extraordinary claim?" Dawkins perhaps adds, "I am so sorry but I can't believe what you say until you supply me with this information." In the second situation, imagine that Dawkins's wife comes home and tells him that she fell from the third floor of a building down to the ground but, miraculously, did not get hurt except for a few bruises. (She even thanked God afterward.) Dawkins, being consistent in his epistemology, says "Great, very interesting indeed. But what evidence do you have that supports this extraordinary claim?" Dawkins perhaps adds, "I am so sorry but I can't believe what you say until you supply me with this information." In fact, imagine that this is also the way Dawkins responds to everything that his friends tell him. My point is that in the scientific scenario, Dawkins's response is just standard procedure; but proceeding in the same way in the second scenario would probably run Dawkins the risk of losing both his wife and his friends.

I am not going to go into all the details of why we assume – for good reasons, I think – that different epistemic norms apply in these cases, but it is sufficient to say that the scenarios illustrate the danger of taking the norms from one (no matter how successful) magisterium and thinking that their application would automatically improve another magisterium. One could also claim, on good grounds, that the context of religion resembles more the context of these everyday life situations than the context of science. Believing in God, at least in the major theistic faiths, is a matter not of mentally assenting to a set of propositions, but of relating to and trusting God, much as we relate to and trust our spouses and friends. Maintaining this does not mean, however, that one could not argue that religious believers sometimes are too dogmatic and uncritical in their religious beliefs. I, for one, think that this is true, and that therefore the epistemic norms of many religious believers could be improved.

Even if Dawkins fails to capture the epistemic norms embedded in the biblical narrative and its rationale, it is still, I think, a mistake to adopt Gould's position of respectful noninterference between the two magistera when it comes to issues of epistemology, the reason being that everything we can learn in one area of life from another area that allows us to improve our cognitive performance ought to be taken into consideration by rational people. We cannot, therefore, exclude the possibility that there could and should be an overlap between science and religion with respect to epistemology or methodology.

But being cautious seems to be a good virtue to cultivate in this context. Certainly, there is nothing in the scientific training that Dawkins and Wilson have received that makes them particularly adept at understanding what religion is all about – understanding, for instance, the kinds of epistemic norms that are or ought to be used in religious practice. If biologists or any other scientists fail to take this into account, their proposed improvements may turn out merely to reflect the imperialistic aims that Gould is afraid drive some of his colleagues. Moreover, even if there could and should be an epistemic overlap between science and religion, biologists such as Wilson or Dawkins have no special authority *as scientists* to suggest epistemic improvements, other than those that concern their own scientific practice. Epistemological evaluations of human practices other than the scientific one are no part of their assignment as evolutionary biologists. It would then be a kind of misuse of science to pretend that a comparison between a religious and a scientific "ethos" (to use Wilson's term) is something that can be made in the name of science.[33] Another kind of misuse of science would be to maintain, as a scientist, that it is not possible to be an intellectually fulfilled religious believer unless scientific epistemic norms can confirm religion. That would be to assume that the only road to truth or rationally justified belief is the scientific path, which may be true or false (probably false) but nonetheless is an issue that does not fall within the scope of the sciences.

V. AREAS OF INQUIRY IN SCIENCE AND RELIGION

Let us turn to the areas of inquiry in science and religion, and to the claims about life and the cosmos that we can find in these practices. Gould advocates respectful noninterference in this area also, whereas Dawkins and Wilson defend the possibility of interference, whether respectful or not. Which position is more reasonable?

According to Gould and the principle of NOMA, each domain frames its own rules and admissible questions and sets its own criteria for judgment and resolution. Science covers the empirical realm and religion the realm of ultimate meaning and moral value, and there should not be any overlap of these magisteria. Gould takes this to mean that "facts and explanations developed under the magisterium of science cannot validate (or deny) the precepts of religion."[34] We can add that beliefs and values developed under the magisterium of religion cannot validate (or deny) the precepts of science.

Nevertheless, Gould has a number of things to say, as we have seen, about the conception of God and of ultimate meaning. On these issues, he maintains

---

[33] Wilson, *On Human Nature*, p. 201.   [34] Gould, *Rocks of Ages*, p. 215.

that the view he defends is similar to Darwin's. Darwin seems to have accepted overall design; Gould quotes him as writing, "I am inclined to look at every-thing as resulting from designed laws, with the details, whether good or bad, left to the working out of what we may call chance."[35] Gould does not explicitly embrace this view; he merely states that the universe, for all we know, may have an ultimate purpose and meaning. Moreover, Gould maintains that this issue cannot be adjudicated within the magisterium of science. But despite this, he thinks that evolutionary theory undermines the idea that there is a reason why we *Homo sapiens* came into being. He claims that life is "a detail in a vast universe not evidently designed for our presence" and that "*Homo sapiens*... ranks as a 'thing so small' in a vast universe, a wildly improbable evolutionary event, and not the nub of universal purpose."[36]

Within the magisterium of religion it is, however, typically believed that the universe and life – in particular, human life – have an ultimate meaning. Jews, Christians, and Muslims, for instance, think that the universe was created by God and that God intended to bring into being creatures made in God's image, creatures like us. They may have come to believe these things by using the sources of knowledge specific to the magisterium of religion – for instance, through divine revelation.

What is puzzling is how Gould *as a scientist* could maintain that a religious belief in the ultimate meaning of human life ought to be rejected and at the same time proclaim that "facts and explanations developed under the magisterium of science cannot validate (or deny) the precepts of religion."[37] These claims are not compatible. Gould tells us that, according to the principle of NOMA, religion is misused when we use it as a control belief *in scientific inquiry*, that is, as a way of restricting the kinds of factual conclusions that scientists are allowed to draw from the data they have access to. But is not evolutionary theory then misused when we, like Gould himself, use it as a control belief *in religious inquiry*, that is, as a way of restricting the kinds of conclusions that religious believers are allowed to draw from the evidence they have access to? It is hard to avoid an affirmative answer to this question. It seems that Gould, the scientist, is dictating what religious people should believe about the ultimate meaning of human life.

If we accept the principle of NOMA, then the same kind of misuse of science appears to flourish in Gould's discussion of miracles. The disciples in the biblical narrative we have discussed had witnessed the crucifixion and death of Jesus (John 19). A few days later, they were gathered together in a

---

[35] Ibid., p. 198.
[37] Ibid., p. 215.

[36] Ibid., p. 205–6.

room, fearing the Jews, when suddenly "Jesus came and stood among them and said, 'Peace be with you!' After he said this, he showed them his hands and side. The disciples were overjoyed when they saw the Lord" (John 20:19–20). On the basis of these experiences, they believed that Jesus had risen from the dead. This belief became a central part of their teaching, and from then on, belief in the resurrection of Jesus Christ became a central part of Christian faith. If we take the biblical narrative at face value, we clearly have a miracle here. Gould writes, however, that

> The first commandment for all versions of NOMA might be summarized by stating: "Thou shalt not mix the magisteria by claming that God directly ordains important events in the history of nature by special interference knowable only through revelation and not accessible to science." In common parlance, we refer to such specific interference as "miracle" – operationally defined as a unique temporary suspension of natural law to reorder the facts of nature by divine fiat. (I know that some people use the word "miracle" in other senses that may not violate NOMA – but I follow the classical definition here.) NOMA does impose this "limitation" on concepts of God. . . .[38]

But how can Gould claim anything like this and at the same time accept the principle of NOMA? Questions about the concept of God and God's actions must by all means belong to the magisterium of religion: what else would otherwise belong to it? What NOMA instead does is to impose restrictions on the use of religious beliefs *within* the magisterium of science. In fact, it forbids any such use. Thus one should not appeal to miracles as a kind of scientific explanation. Scientific inquiry is restricted for methodological reasons to empirical explanations, not a priori but because such a restriction has proved to be successful. But that constraint puts no restriction on whether religious believers *within* the magisterium of religion should believe in miracles. According to NOMA, that is a topic for religious inquiry.

Thus, Gould cannot consistently claim that NOMA imposes limitations on concepts of God, of miracles, or of the ultimate meaning of life within the magisterium of religion. Consequently, the first commandment of NOMA is not: "Thou shalt not mix the magisteria by claiming that God directly ordains important events in the history of nature by special interference knowable only through revelation and not accessible to science." It is rather: "Thou shalt not mix the magisteria by claiming that the belief that God directly ordains important events in the history of nature ought to guide or restrict scientific inquiry."

[38] Ibid., p. 84–5.

I have argued that Gould's account is inconsistent. But is it not reasonable to take, for instance, the things he says about the relevance of evolutionary theory for questions of meaning as an indication that it is, in fact, the principle of NOMA that we ought to reject? We would then accept that the facts and explanations developed under the magisterium of science can validate or undermine the precepts of religion. To some extent, I think this is correct. But we must be careful not to obscure the issue by conflating science with scientism.[39]

Gould writes, as we have seen, that "*Homo sapiens*... ranks as a 'thing so small' in a vast universe, a wildly improbable evolutionary event, and not the nub of universal purpose."[40] The idea seems to be that all biological events taking place in evolutionary history, including the emergence of our species, are random with respect to what evolutionary theory can either predict or retrospectively explain. Therefore, there is no ultimate meaning to human life. Humans are not planned by God or by anything like God.

Perhaps it is true that the existence of human beings is a wildly improbable event given the information that is accessible to scientists through the use of biological methods, but how can we, from this information alone, conclude that we are not intended by a God to be here? It seems that we need an extra premise saying that the scientific account is exhaustive: what science cannot discover does not exist, or at least we cannot know anything about it. If evolutionary theory implies that our existence is a widely improbable event, and if *the only source of knowledge we have is science (or more specifically, in this case, evolutionary theory)*, then it follows that we ought to believe that our existence is the result of pure chance. But that is to conflate science with scientism.

The relevant issue for the magisterium of religion, however, is not what is likely given the theories that scientists possess, but what is likely given what the practitioners of religion take *God*'s knowledge to be about the outcome of the evolutionary processes. The participants of this magisterium may disagree about the extent of such divine knowledge, but they (if we have Jews, Christians, and Muslims in mind) typically have in common a belief that God's cognitive capacity outruns our capacity by far. Thus, God's ability to predict with great accuracy the outcome of future natural causes and events

---

[39] Scientism is, roughly, the view that the only things that exist (or that it is reasonable for us to believe exist) are the ones science can discover and that the only kind of knowledge we can have is scientific knowledge. See Mikael Stenmark, *Scientism: Science, Ethics and Religion* (Aldershot: Ashgate, 2001), Chapter 1, for a discussion of different forms of scientism.

[40] Gould, *Rocks of Ages*, p. 206.

is probably enormous. We cannot, therefore, automatically assume that what is likely given such vast knowledge is the same as what is likely given what evolutionary theory can predict or retrospectively explain. So if God planned to create us and if it is likely that we would actually come into existence, given what God can know about the future of the evolving creation, then one could reasonably claim that we are here for a reason, and that there is a purpose, in this sense, to our existence. To establish the opposite conclusion requires more than basing one's calculation of probable outcomes on the current version of evolutionary theory. It follows that a successful argument for this conclusion takes us outside the domain of science and into metaphysics and theology. Hence, Gould's and other scientists' inference from evolutionary biology that human existence is purposeless cannot be categorized as scientific.

Although it cannot be demonstrated in this context,[41] it is exactly this that is problematic with the attempts made by Dawkins, Wilson, and others to expand the boundaries of contemporary evolutionary biology into the fields of ethics and religion: they tend to conflate science with scientism. I would, therefore, reject not only Gould's independence view of science and religion but also Dawkins's and Wilson's conflict view and maintain instead that the most reasonable position to adopt is some kind of contact view. *Pace* Gould and the principle of NOMA, we ought to accept that there can be an overlap between science and religion not only on the social but also on the method-ological and theoretical levels. Contemporary biologists, on such a view, are Darwinian heretics when they let their prior ideological or metaphysical com-mitments determine what implications evolutionary theory has for religion, morality, or human life in general. In short, Darwinians ought not to confuse science with scientism.

---

[41] See Stenmark, *Scientism*, for such an argument.

Index

www.ingramcontent.com/pod-product-compliance
Ingram Content Group UK Ltd.
Pitfield, Milton Keynes, MK11 3LW, UK
UKHW040705180125
453697UK00010B/427